KW-621-525

1 (*frontispiece, overleaf*) Butty boat helm

The Canals of Britain

D. D. Gladwin

With illustrations by J. K. Ebblewhite

B. T. Batsford Ltd
London

First published 1973
© D. D. Gladwin, 1973

ISBN: 0 7134 0492 2

Printed in Great Britain by
Cox & Wyman Ltd, Fakenham
for the publishers,
B. T. Batsford Ltd
4 Fitzhardinge Street, London W1H 0AH

Contents

The Plates

Acknowledgements

I am greatly indebted to the many friends, colleagues, acquaintances and professional bodies who have been kind enough to provide material and photographs for use in this book. The omission of their names is not intended to be in any way derogatory to their efforts and my thanks are none the less sincere. Among these and deserving of special mention are: Hugh Barker, George Bate, B.E.M., Arthur Cross, Martyn Denney, L. A. (Teddy) Edwards, Phil Garrett, A. J. Harmsworth, G. Harris, B. Holding, Jack James, F. W. Jones, D. G. Russell, Peter L. Smith, N. L. Thomas, Alf Watton, Captain A. V. West; members of the British Waterways Board, including among others Miss S. Doeg (*Waterways News*), J. Beer (Press Officer), the Section Inspector, foremen and workmen of the Worcester & Stratford section; librarians and staff of the Bromsgrove, Banwell (Som.), City of Lincoln, University of Birmingham, University College of Wales and Worcester libraries; staff at the Waterways Research Centre and Somerset, Southampton, Essex and Berkshire Record Offices; the editors of *Sea Breezes, Marple Rate Payer* and *Record* (Transport & General Workers Union); the Hon. Secretary and members of the Inland Waterways Protection Society and Canal Transport Marketing Board; J. K. Ebblewhite must be singled out for his forbearance and understanding in the provision of scraperboard drawings; and finally my wife, co-author in all but name, typist and general coffee-maker. Hers was not an enviable lot.

The jacket illustration shows a pair of working narrow boats descending the Itchington locks south of Warwick. Photograph by Hugh McKnight.

Photographs have been provided by the following: R. Bird, 41; British Waterways Board (*Waterways News*), 25, 28, 38, 44; M. Denney, 58; L. A. Edwards, 19, 20, 26, 47; P. Garrett, 29, 30, 32, 40; A. J. Harmsworth, 36, 37; G. Quartley, 7; D. G. Russell, 9, 16, 48; W. H. Scriven, 55; S. J. Sherry, 18; P. L. Smith, 15, 27, 34, 35, 43, 49, 57; J. Thornett, 21; Transport & General Workers Union (*Record*), 56; D. Valentine, 39; Waterways Research Centre, 31. For the balance my wife and I must take responsibility. Figure 10 is reproduced by arrangement with British Waterways Board. Figures 16–19 are based upon Ordnance Survey maps with the sanction of the Controller of Her Majesty's Stationery Office, Crown copyright reserved.

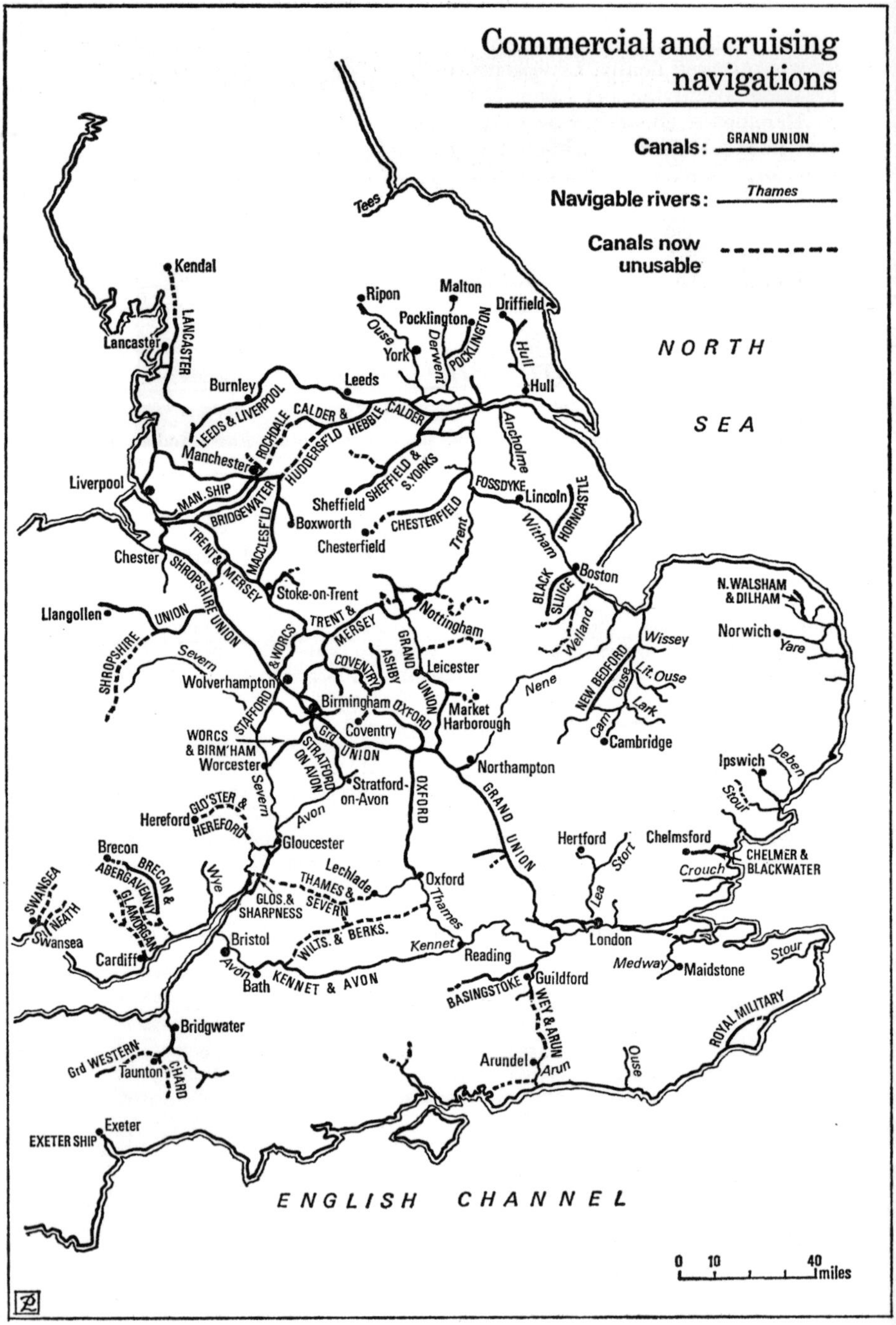
Commercial and cruising navigations
Canals: GRAND UNION
Navigable rivers: Thames
Canals now unusable
NORTH
SEA
ENGLISH CHANNEL
0 10 40 miles
Tees
Kendal
LANCASTER
Lancaster
Ripon
Malton
Driffield
Pocklington
POCKLINGTON
Ouse
York
Derwent
Hull
Hull
Burnley
Leeds
LEEDS & LIVERPOOL
CALDER &
HEBBLE
CALDER
ROCHDALE
HUDDERSF'LD
Manchester
Liverpool
MAN. SHIP
BRIDGEWATER
Ancholme
SHEFFIELD & S.YORKS
Sheffield
Boxworth
CHESTERFIELD
Chesterfield
FOSSDYKE
Lincoln
HORNCASTLE
Witham
Trent
MACCLESF'LD
Chester
TRENT & MERSEY
SHROPSHIRE UNION
Stoke-on-Trent
Boston
BLACK SLUICE
N. WALSHAM & DILHAM
Llangollen
UNION
TRENT & MERSEY
Nottingham
Welland
Norwich
Yare
Wissey
SHROPSHIRE
Severn
& WORCS
GRAND UNION
ASHBY
COVENTRY
Leicester
Lit. Ouse
NEW BEDFORD
Ouse
Nene
Lark
Wolverhampton
STAFFORD
Birmingham
OXFORD
Market Harborough
Cam
Cambridge
WORCS & BIRM'HAM
Coventry
Grd UNION
Worcester
STRATFORD ON AVON
Stratford-on-Avon
Northampton
Ipswich
Deben
Stour
Severn
GLO'STER & HEREFORD
Hereford
Avon
OXFORD
GRAND UNION
Gloucester
Hertford
Stort
Chelmsford
CHELMER & BLACKWATER
Crouch
Brecon
BRECON & ABERGAVENNY
GLAMORGAN
Wye
Lechlade
THAMES & SEVERN
Oxford
Thames
Lea
SWANSEA
NEATH
GLOS.& SHARPNESS
Swansea
Cardiff
Bristol
WILTS. & BERKS.
Kennet
Reading
London
Medway
Maidstone
Stour
Avon
Bath
KENNET & AVON
BASINGSTOKE
Guildford
WEY & ARUN
ROYAL MILITARY
Bridgwater
Grd WESTERN
Taunton
CHARD
Arundel
Arun
Ouse
Exeter
EXETER SHIP

Prologue

To imagine the coming of waterways, their meaning and their impact, it is necessary to forget the non-stop thrum of traffic outside your door, to forget the television louring in the corner, to remember there was no electric light, but at best a candle and at worst a stinking, smoky rushlight, to remember the typewriter was many years away and that thousands of men – not women – painfully wrote out, often in beautiful copperplate, all the details of a contract – and then wrote a duplicate for the file. We must also remember that these men wrote phonetically for English was not standardised. In 1730, the date of the first Act for canalising the Stroudwater, the Jacobite Rebellion had yet to take place in Scotland, the Irish were even then a problem and the Welsh were still proud to be a distinct people. There was no Ministry of Social – or State – Security to look after you when you were sick or unemployed, and the king was a king, not merely a puppet, choosing his government and openly corruptible. George I could not speak English, George II banned his wife from his coronation and George III went a bit odd – but the canals were dug during the reign of all three.

Most important of all at the commencement of the canal era everyone knew his place in the pecking order of society. It is true one could, with a modicum of luck, and more than a dash of influence, work up the ladder. One could, but, and it is a big but, this was a society for men only. Men, whether navvies or gents, ruled the roost, women being for the home.

Still, for all the lack of sewerage facilities, and the poverty, it was a vibrant age, an age of evolution and change.

These are the facts of the background to the building of the greatest waterways of England, Wales and Scotland. Whether canals should remain moribund, advance or die is a matter of opinion and although the story of waterways is given in the following pages together with a few personal opinions, each individual must interpret the facts in the light of his or her own experiences.

I *The Specification of a Canal*

DIMENSIONS OF THE CANAL. The breadth of the canal at bottom is 16′ and at the level of the top of the towpath 42′, the depth of the water 5′, the height of the towing-path and the top of the opposite bank 1′ 6″ above the water surface; the breadth of the towing-path in all cases . . . to be 10′ except under the bridges.

BRIDGES. There are 4 Turnpike-road bridges, 24′ between the parapit, 10 Parish and Bye-road Bridges, 16′ between the parapit, 33 Accommodation Bridges, 12′ between the parapit, 2 Junction Bridges one at each end of the Canal.

AQUEDUCTS. There are 10 Aqueducts upon this branch of Canal from 6–15′ span; and one of 3 arches of 10′ span each.

LOCKS, ETC. There are to be 23 Locks, 5 Lets off, 5 Sets of Drop Planks, 7 Waste Weirs, 45 Single Culverts from 2–4′ diameter; besides 5 double two-feet and 7 single two-feet Culverts to be made of English Oak free of sap, for the Morasses &c. &c.

The whole of the Works to be completed in a Substantial and Proper Manner to the satisfaction of MR. TELFORD, the Engineer in Chief, within two years after the signing of the

> Contract, unless prevented by some accident of circumstances over which the contractor could have no controul.[1]

A canal is a physically complex structure. If one looks at any area of land, whether on a map at home or when out driving, and imagines cutting a 42-ft-wide ditch through it, all on one level, solely by the muscle power of men and horses, this much will be quickly apparent

An ordinary field of four acres will almost certainly have a drainage ditch which must be culverted to pass under the canal, or have to be diverted away. Trees, hedges and undergrowth must be uprooted and burned. A barn, stable or hut will need to be rebuilt, the farmer's right of access to his other lands may well need to be protected by an accommodation bridge and the bifurcated field fenced to prevent animals straying. All these triviata must be planned even as the necessity of a cutting or embankment is considered. And these particulars must all be specified, even if not in detail, by the surveyor and his assistant before plans are laid in front of Parliament. And worse, permission must be granted by the farmer or landowner, for if he disagrees it may be that some vast subsequent diversion must be made.

In 1819 Thomas Telford said, and every subsequent engineer of the Birmingham & Liverpool Junction Canal has echoed:

> In passing through Cheshire, the canal was . . . peculiarly unfortunate, for the marly soil, of which the surface of that rich country chiefly consists, when used for embankments slips and bulges in great masses, and rapidly dissolves when exposed to the atmosphere. From an unaccommodating disposition in some of the landowners, in persisting to prohibit the proper line of canal, these evils were experienced to an enormous and unprecedented extent . . . [2]

The landowners' influence at that time is a contrast with the methods now in force; where motorways – a roughly parallel feat – are concerned, nothing seems to be sacred.

Having, one way and another, drawn up a line of the waterway, and the Act for the building passed – details of which will be found in chapter 2 – and some, if not all of the money raised, the work would then be put out to contract utilising small contractors or, often, direct labour gathered for use under the eye of the engineer. The problems for the later canals of shortages of good labour were as bad as they are today, and it is advisable at this juncture to take a sideways look at the politico-economic situation in Great Britain in 1760.

At the time when the Duke of Bridgewater was cogitating on his plans for a canal, the life of the working man went on much as it had done for the last hundred or so years. To walk to the market town – perhaps not more than ten miles away – with the wife and children suitably and relatively well-dressed for the fair day, was the equivalent of our going to Majorca. The strange places, strange faces and strange accents heard! The beer was better, the food sweeter and sun warmer on this day.

The labourer's cottage – and here we refer to a hard-working man in a reasonably affluent part of the country – was small, somewhat lacking, to our eyes, in the sewage department but warm and comfortable. Hire-purchase facilities existed, of course, and were a necessity; typically, boots were made and worn in the spring, payment, it was understood by both cobbler and wearer, being made from the harvest bonus. 'In the sty a pig, in the wife a child, in the furrow a seed' was an apposite description of these men's existence. While the price of flour and beer remained stable against the wages there was little to complain about, things were no better or worse than in their father's day, and he hadn't grumbled . . . True, there were rumblings from some discontented men, but these were little heeded by the mass.

Coalmining was still incredibly primitive, for the basic method was to dig hundreds of little pits in the ground, each with minuscule tunnels leading off. Although many of the early river navigation Acts were centred on mining areas, there was little incentive for a man like Scroop, the Duke of Bridgewater's father, to extend his pits when there were no means, save carts in summer, of shifting the coal to the consumer. Even presuming a dry summer and a modicum of luck, only the best paid of men could afford to buy the fuel, the remaining 80 per cent of the population, living as they did in rural areas, contented themselves with traditional materials – wood, peat or dung.

Corn would be grown in the fields of the villages and milled by the miller. It has been estimated that nowhere in England could one, at that time, be further than ten miles from a water- or wind-mill; a day's journey for a carter. Ironworkers, logically, lived either near the source of iron or of wood, and at least within a reasonable distance of the navigable rivers. The few domestic utensils required by a labourer could be bought from a travelling tinker, or, more probably, from the blacksmith. Bricks and pottery were likewise produced on a day-to-day basis. What little pottery the average householder possessed would be brought by the monthly travelling cart, a superior form of tinker who carried everything from Redditch needles to Scotch yarn. It is noticeable that, excepting the houses of the gentry, most red-brick dwellings were to be found near the brickfields or near a navigable waterway. Napton village and Stratford (with the Avon as its navigation) are good examples. In this context one should not belie the efforts and effects of coastal sailing craft, which provided an excellent intercourse between industrial areas centred around the Trent and what was to become Cobbett's 'Great Wen', London; even by 1750 the latter represented a parasitical growth, inasmuch that very little of its produce was home manufactured. In remoter parts, such as Dorset, mud dwellings were still being erected; elsewhere wood was used.

The cities were growing; a greater import/export movement meant an increase in white-collar staff and their attendant families concentrated near the great ports and cities. The population figures for Manchester and Salford show this rate of growth all too clearly:

1757	*1773*	*1821*	*1824*
19,800 *est.*	27,250 *est.*	133,788 *census*	164,000 *est.*

While the population of England and Wales as a whole also increased, but not to the same proportions:

1688	*1760*	*1801*	*1811*	*1821*
5·5 m.	6·5 m.	8·9 m.	10·2 m.	12·0 m.

2 A river navigation. Keynsham lock and weir on the Somerset Avon

3 A hire boat on the Shropshire Union Canal, Welsh Section

4 Horse-drawn days. Note wear on the right-hand iron guard and bri
girder. Birmingham Canal Navigations, 1966

5 Dutton, St Helen's Canal, 1840. Drawn to show the contrast between t
antique lock and the modern railway

Principal imports also increased:[3]

Decade	*Corn* £,000	*Groceries* £,000	*Total Imports* £,000
1750–59	40 (–)*	2,641 (32)	8,325
1760–69	226 (2)	3,863 (36)	10,719
1770–79	421 (3)	3,918 (32)	12,104
1780–89	422 (3)	4,295 (31)	13,820
1790–1800	1,168 (5)	7,633 (35)	21,797

* The figures in brackets are percentages of total imports.

The building of the Duke of Bridgewater's Canal is often cited as a masterpiece:

> This then was the first step taken towards making this very early and useful navigation, but the degree of supineness exhibited by the original undertakers in so long neglecting the execution of the work which has been, and is yet, a source of immense wealth, is most strikingly contrasted by the enterprising spirit and the astonishing perseverance of the Duke of Bridgewater who unassisted, except by the natural genius of Brindley, carried into execution a series of difficult and expensive works which are even at this time unexampled.[4] The general design of these works is, undoubtedly, great, the whole plan shews a capacity and extent of mind which forsees difficulties, and invents a remedy before the evil exists . . .[5]

But in fact, it was only a part of a natural trend.

Granted, in many ways, it was an improvement on its forebears, of which the Warwickshire Avon and the Stour are examples, neither being entirely successful for either the proprietors or the population served, while the Duke made a good profit. The greatest strength of the Bridgewater Canal lay not only in the ability of Brindley to combine aqueducts and a level line with a wide canal – one where barges of an economic size could be used – but also in the concealed waterways that ran from Bridgewater's mines.

> . . . We pushed up this subterraneous passage (just wide enough for the boat) above a mile . . . other passages lead from this principal one into divers parts of the mountain and recesses cut in the rock suffer one boat to pass another. In this singular voyage it is almost impossible to believe one's self in motion; the rocky passage, arched with brick, seem to be flying from you, and makes the head dizzy, and in returning the distant entrance looks like a bright star.[6]

Pretty awesome! One finds it difficult to visualise a dark, groaning, gaseous hole in the earth, where the sole illuminant was a candle. But men worked there – and with a 14-hour day, nigh on lived there.

Having won the coal and placed it into boats there were two ways of transhipping it into the waiting barge – by hand with long shovels that could lift a hundredweight at a time, or by the forerunner of our modern containers, wide boxes with iron straps holding two tons a piece, which were then craned across. Thus far was Brindley a genius, but a mystery that has never been solved, and which bears heavily upon waterways now, is why he should have built the Bridgewater, Derby and Droitwich barge canals of a wide (i.e. in excess of 14 ft) beam, and his trunk canals to a narrow (i.e. 8 ft or under) beam. The long-term legacy of these narrow canals is simply that they are outclassed today and even where a wide canal exists the connecting links are often narrow. The illogicality of the whole thing is even more apparent when one considers that the majority of waterways were built by small, independent, companies with different engineers, any one of whom, it being a British characteristic, could have been bloody-minded and set the trend for a wider canal. This is especially true when one takes the example of the Birmingham & Liverpool Junction Canal whose engineer, Thomas Telford, had already been engaged on the widening works on the Birmingham Canal Navigations Main Line, providing a canal with dual towpaths and superb tunnels, and yet he built the Birmingham & Liverpool Junction not only with narrow locks but with narrow cuttings only capable of taking a single boat. Granted there was some saving in cost both in land purchase and material cost in building a narrow waterway, but in the heyday of waterways, until the coming of the rail-road, this was not a matter of any great importance; having heard of the spectacular dividends paid on the Trent & Mersey Canal it is improbable that shopkeepers or clerks with dreams of making hundreds of pounds overnight would have cared whether the engineer quoted £28,000 6*s.* 2½*d.* or £48,000 6*s.* 2½*d.*

An even more curious fact is that some canals were built part wide and part narrow. This is the case on the Trent & Mersey – a narrow canal between two (wide) rivers.

> Harecastle [old] Tunnel, upon this canal was constructed by James Brindley for a distance of 2,888 yards, at a level of 197 feet from the highest summit of the hill above it. This tunnel would only permit a 7 feet boat, with a moderate lading, to pass through it, and then only by employing leggers to propel it, a class of men who, lying upon their backs on the freight, pushed against the sides and top of the roof with their feet, and thus moved the boat onwards. The tunnel is only 9 feet in width, and 12 feet in height, and occupied nearly eleven years to complete; and to pass a boat through it occupied two hours. In the year 1824 Thomas Telford commenced another tunnel, at a distance of 26 yards from that already described; its length of 2,926 yards, its width 14 feet, and height 16 feet, of which 4 feet 9 inches is covered by the towing path, leaving 9 feet 3 inches for the passage of the boat; the path is supported by pillars, and the water flows under it. There are altogether 15 shafts sunk, well set out, and the headings so accurately driven, that the whole length may be seen at one view.[7]

This was not an inexpensive operation, costing in all some £112,681.

The care and attention that was required by the engineer and contractor is apparent even in such a small detail as the specification of bricks and mortar. 'The bricks were made from clay of superior quality, after being triturated by machinery; the mortar was of Barrow lias limestone, ground in a mill, and when set is quite impervious to water.'[7]

Sporadically waterways have made an attempt to improve their service by widening the cuttings, eliminating the odd tunnel and, of course, by dredging. In the case of the Northern Oxford at Hillmorton a new set of locks was installed virtually alongside the existing ones so that now the British Waterways Board have two radically different sets of paddle-gearing to maintain. In the 1930s, with a little financial assistance from the government, the Grand Union Canal Company embarked upon an adventurous scheme to widen the waterway from London to Birmingham. In this case a grant was given by the government to relieve unemployment, and the old single 7-ft locks were replaced by 14-ft models with up-to-date shrouded paddle gearing; but not only was there no attempt to eliminate the unnecessary kinks in the waterway, the tunnels and bridgeholes and even the passing places were left as they were so that only at the big flights, Hatton, Itchington, etc., was there any real advantage to be gained when using a motor boat and butty – the two could work together, side by side, rather than the butty having to be hauled behind the motor.

The basic design of a lock does not materially alter whether it be capable of taking craft of 7-ft beam, 14-ft beam or even the big barge locks of the Gloucester & Berkeley Canal (90-ft beam), but the details have altered – or evolved – over the years. The latest style is in use on the Caledonian Canal where the work of conversion began in 1959 when a pilot scheme, involving the electrification of existing equipment, was tried out at Clachnaharry. Although successful, it was felt that the method could be bettered and the present very efficient electro-hydraulic system of operation was evolved. Provision has been made for a system of standby pumping to enable the locks to be operated, although at a slower rate, in the event of a mains power failure.

Old fashioned, manumatic locks comprise a number of relatively simple units. Firstly, the basic stone or brickwork is itself an object of more devolution than evolution, for old age has too often eroded the sides into a concave shape. One cannot say though that the masonry of a lock will necessarily be brick and stone, other than as a generalisation, for Beeston top lock is made of cast iron plates riveted together, as it was found that the lock was positioned neatly in green sand and refused to stay in shape. The River Kennet has turf-sided locks, while a lock on the Caledonian Canal is cut from the living rock with no more than a facing of brickwork. Again, that weird-looking contraption at Thurlwood on the Trent & Mersey Canal, built to combat subsidence, is like no other lock in Britain. Disregarding such exceptions, neither the sides nor the bottom of a lock are necessarily parallel or flat, more often than not there will be gullies in the base brick or stonework enabling the water, when the lock is dewatered for maintenance purposes, to drain into the paddle-holes from the top end. The sill against which the gate abuts will almost certainly be a solid lump of elm standing proud of the bottom. The stonework into the chamber of the lock, if it is a very early canal, will be dowelled one stone

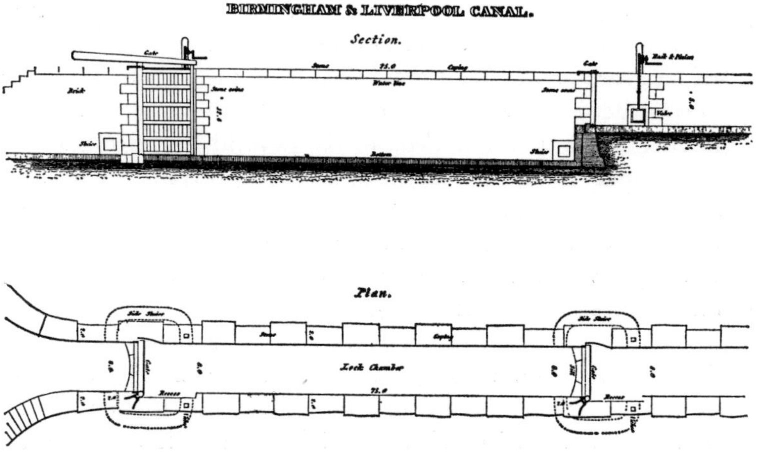

Fig. 1. Canal lock, by Thomas Telford, c.*1825.*

to another with iron or lead for the dowels. The part against which the boats' fore-ends will rub will again be of elm bolted with inch iron bolts, occasionally direct into the stonework, but usually to a plate set in the base sill. The sides of the chamber are often battered (too often, in both senses!) and it was stated by an engineer, Mr Cresy, in 1833 that:

It is by no means advisable to make the walls of the lock chamber strictly perpendicular; some inclination, or talus, is always preferable, particularly at the back, whence resistance is required against the thrust of water; and as its filtration is another point ofimportance to be guarded against; a greater thickness is necessary at the bottom than at the top of the wall, that filtration being greater in proportion to its height. The walls should, however, be perpendicular towards the interior of the lock, at least in that part where the boats mount and descend; the lower part may have the same talus as that given to the boats. The least thickness for the walls at the level of the water is 4 feet 3 inches, in order to allow for a course of beton, as a protection against the filtration of water: as a general rule, the thickness should be equal to half their height, and as the walls become more lofty, their foundations must be spread out in proportion. Inverted arches have been introduced with considerable advantage to form the profile of the wall of lock chambers; where there is a good foundation they may commence with a considerable batter or inclination, rising gradually in a curved form, finishing at the top with a perpendicular face; such a wall resembles an arch, from its having the joints and beds of the several stones made perpendicular to their face, which also prevents them from sliding on each other; the triangle of earth, supported by the back of the upright retaining wall, may be regarded as a wedge, of which its own gravity is the acting power, and it is only supported by the re-action of the inclined plane of earth, and the back of the wall, each of which acts perpendicularly to its plane. The breadth of such retaining wall must be always proportionate to the height, and to be of equal specific gravity with the earth, its breadth should be more than $\frac{1}{3}$rd of its height. The specific gravity of walls built of brick or stone is a trifle more than that of the earth they support, therefore the dimensions may be slightly diminished.[7]

Some detail of the fitting of the gates will be given in chapter 7, but taking a standard narrow lock with a single gate and double (so-called mitre) bottom gates, the drawing on p. 23 gives more of the basic structure, but even here on a perfectly ordinary canal differences abound. On the Northern Stratford, possessed of 19 operative locks, two have lower gates steel-framed as a wartime economy measure. On the Wolverhampton flight of 21 locks, the second up from the bottom has a single top gate both at top and bottom. As first built the canal had a very deep bottom lock which was found to be extremely wasteful in water and sometime in the 1830s this was replaced by a pair of locks, the new 20th lock being built differently as the engineer at that time thought differently to his predecessor. The basic reason for fitting double gates at the bottom end is that double gates are capable of withstanding a greater pressure of water than a single gate. The weight upon them forces them back into the quoin and against themselves. It should be mentioned in this context that aside from the thumping of a boat against a lock gate, a full standard single lock of 8 ft drop (7 ft × 70 ft) contains something in the order of 40,000 gallons of water, a not inconsiderable static pressure being exerted by such a quantity. Particular note should be made of the anchor to which the collar strap is attached. Until quite recently it was a rarity to find these set in anything less than a half-ton slab of sandstone or granite. The anchor arms are clawed to a depth of a foot and bonded to the stone with molten lead. About 4 lb. of lead are to be found attached to each claw of the anchor. Fundamentally, as can be seen, there are only two items which hold a lock gate in position, the first being the collar, which in the past was invariably greased to add to the ease of swinging the gate, and the heel-post which operates in a ball and socket joint, the correct name for the parts being a box and a pivot. One of the difficulties past and present is that of keeping these gates in alignment and this is often the only clue that will be given when a lock starts to 'come in'. This, fortunately rare, occurrence may result in the lock sides closing at a rate of up to 2 in. a year.

The apron at the bottom end of the lock is normally constructed of brick, as are the wing walls. The position of the steps used for access to and from boats varies from lock to lock, the design being modified to suit local conditions, especially in cases where a crossover bridge is in close juxtaposition to the bottom gates. Incidentally, it is interesting to note that with the increase in road traffic some of these crossover bridges have been widened and it can be necessary to fit very short, or cranked, balance beams on the bottom gates. Good examples of these are to be found at York Street and Kidderminster Town Lock on the Staffordshire & Worcestershire Canal, where what were once rural lanes are now busy highways.

Bridges can be divided into a number of distinct groups. The first are accommodation bridges. The best known of these are the red brick bridges of the Midlands. Although declining in number now, the red bricks have weathered well for 160 years, however age and excessive weight is taking its toll and regrettably where they are replaced it is more often than not by some featureless concrete structure, a good example of which is on the Grand Union Canal at Itchington, possibly one of the ugliest but (one hopes) economic to build.

Swing-bridges abound in the flatter parts of the countryside, on the New Junction Canal, although operated manually by chain and pulley they are functional and impressive in their setting. Older and much photographed, due to their proximity to the lock-keepers' 'Georgian' cottages, are those on the Gloucester & Berkeley Canal.

Lifting bridges which vary enormously in size and in manner of operation, have their own problems. On the Caldon Canal the bascule type is utilised, being counterbalanced by weights in a box at the top. One bridge had always given much trouble to a local farmer who, finding that it moved under his tractor, removed the iron ballast, thus ensuring that it stayed down! Those on the Welsh Canal have been much photographed, while the twin-arm type on the Southern Oxford is again well known. Some pleasure is given to boats' crews when cruising the Leeds & Liverpool Canal, for when the bridge at Litherland has to be swung, it disturbs the passage of cars on a busy highway and quite considerable queues can build up, the expressions of the drivers, to borrow a phrase, are 'wondrous to behold', but most of these bridges, fortunately, tend to be in quiet rural areas. Mention should be made here of one very impressive modern swing-bridge, that over the docks at Bristol. Electrically operated, it is controlled from a crow's nest and with an overall weight of over 100 tons swings as easily and gently as a matchstick on a pond.

Crossover, or junction, bridges are accounted for by going back to the days of horse-drawn craft, when the towpath had to change sides. That near the Blockhouse at Worcester on the Worcester & Birmingham Canal has ceased to bear any resemblance to its original design, being now a four-lane highway; any horse venturing to cross would rapidly be converted into dogs' meat. Others, as at Bollington on the Macclesfield Canal, look forlorn with the grass growing high between the parapets. The curves of the parapet wall leading to and from these bridges are often graceful but, regrettably, unappreciated. The reasons for the towpath changing over are three. The first, where it is necessary for the towing-path to cross over an arm of the canal or at the junction of two canals. The second, where having met with difficulties with landowners, the only way the towpath could be built was to construct it on the opposite side of the canal. The landowners lived in fear of boatmen poaching if they had easy access to the fields! The third, which sounds remarkably humane, was supposed to be that in conjunction with a slightly cambered towpath, if the horse had to cross over every few miles it evened out the wear on its shoulders from the towline.

Some of the most attractive bridges are to be found at the tail of a lock. Often in cast iron, they may be there for the purpose of permitting the boatman to run from one side of the lock to the other without crossing the lock gates, alternatively they may be there to protect an ancient right-of-way and, indeed, may often now be the only clue to the existence of such. At lock 36 on the Worcester & Birmingham Canal, known as the Clap Bridge, is an enormous slab of stone serving as a bridge. This differs from most, inasmuch as it is not split in the centre as are those on the Stratford and Staffordshire & Worcestershire Canals, but is offset to one side.

A strong case may be brought forward for the retention of a number of cast and wrought iron bridges if only for their vintage – and fitting – appearance. Some efforts are being made towards this and at Braunston on the Grand Union Canal, where there is a superb double roving bridge over both the arm that once led to Barlow's Yard and over the Oxford Canal, one is pleased to see it being well kept up. Others, plainer, are still to be found on the Birmingham Canal Navigations, often marked with both the name of the iron works and the casting date. Regrettably, a number of bridges were knocked down to aid the wartime scrap drive, others having been left derelict when the canal was closed and many neglected ones are bound to fall down eventually. This is an unnecessary form of vandalism, if a beautiful thing exists surely it is better to retain it than let it go by default. Whilst on the subject of preservation, or conservation, one must mention the humble accommodation bridge which is also suffering from vandalisation from both children and mad motorists, for upon repair work being carried out no attempt is made to match the indigenous materials, any old 'bodge', seemingly, sufficing. There is no better way of showing the intelligent use of natural local materials than to look at the handsome high-backed accommodation bridges of the Pocklington Canal or the granite of the Cromford Canal or the Bath stone which abounds in Somerset.

Whilst most bridges are plain, nevertheless in certain circumstances it was necessary for the engineer to pander to the landowners' ideas and to build a bridge which was more pretty than functional. The very nature of a canal, fortunately, made for some restraint. Ladies Bridge on the Kennet & Avon Canal is a more or less normal bridge with fine examples of a stonemason's craft and Avenue Bridge on the Shropshire Union Canal Main Line is merely an example of the architectural style fashionable at that time. This restraint, however, is not apparent where Drayton Manor footbridge on the Birmingham & Fazeley Canal is concerned. It is a superb example of the nonsensical – nothing more or less than the castle's portcullis, absentmindedly misplaced by Richard Coeur de Lion.

There can be little doubt that motorway engineers are often forced to use considerable ingenuity when providing their accommodation bridges over a canal, and it is not unreasonable to point out that if we had modern canals these motorways could have been kept in their place!

In the specification of the Newport Canal – given at the beginning of this chapter – aqueducts will be seen to play a relatively small part. Best known of all aqueducts is, of course, Pont-y-Cysyllte on the Welsh Branch of the Shropshire Union

Canal; built in 1795 it is unlikely that anything in the engineering field will ever be made to equal it.

The Aqueduct of Pont-y-Cysyllte is 4 miles from Chirk, over the river Dee; an earthen embankment was pushed forward until the perpendicular height became 75 feet. The distance between the end of this and the north bank is 1,007 feet, and the river Dee is 127 feet below the water level of the canal carried over it. To construct an aqueduct upon the usual principles, with piers and arches, 100 feet in height, and of a sufficient breadth and strength to afford room for a puddled water-way, would have been, not only extremely hazardous, but expensive. Telford, who had already carried the Shrewsbury canal by a cast-iron trough 16 feet above the level of the ground, formed the idea of doing the same in the present instance, and made a model of a portion set on two piers, with the towing-path and side rails, which was approved of and finally adopted. The foundation on which the piers are erected is a hard sandstone rock; their height above low water in the river is 121 feet, at the bottom they are 20 feet by 12, and at the top 13 feet by 7 feet 6 inches. For a height of 70 feet from the foundations, they are built solid, and the remaining 50 feet hollow, the walls being only 2 feet in thickness, with one cross inner wall; by this means the centre of gravity is thrown lower in the pier, and the masonry economised. The width of the water-way is 11 feet 10 inches, of which the towing-path covers 4 feet 8 inches, leaving 7 feet 2 inches for the boat; as the towing-path stands upon iron pillars, the water fluctuates and recedes freely as the boat passes. There are eighteen of these stone piers, besides those of the abutments, and the total expense, including the embankment, was £47,018.

	£	*s.*	*d.*
The embankment cost	8,570.	15.	8
Masonry	21,162.	13.	5½
Iron work	17,284.	17.	5½
	£47,018.	6.	7

This aqueduct almost rivals the works of a similar kind left us by the Romans.[7]

Equally well known, howbeit constructed of different materials, are Dundas Aqueduct on the Kennet & Avon, Marple Aqueduct on the Peak Forest and Almond Aqueduct carrying the Union Canal in Scotland; but how rarely people notice the little ones. Driving under in a car, or even over in a boat, it is difficult to visualise how much of a problem it is to retain a canal in a channel 20, 30 or 40 feet up, and different engineers met the problem in different ways. Some were successful, Brindley's aqueducts carrying the Staffordshire & Worcestershire Canal over the Sow are cumbersome but solid and seem to bulldoze their way across the countryside while Bevan's on the Grand Union Canal at Itchington is hardly recognisable as an aqueduct from the canal bank, but with a river that has a rise of 12 feet between summer and winter levels underneath, it has to be strong. The new aqueduct at Ryland Road on the Birmingham Canal Navigations New Main Line presented modern engineers with some problems, how then must Telford – by then an old man – have felt as he passed the drawings for the aqueduct at High Onn on the Shroppie Canal?

Locks we have already mentioned but those mysterious 'lets-off' in the specification need some explanation. A problem often faced by engineers is what to do with waste water in winter time when the drainage from the fields, roads and houses overfills the canal. At certain strategic points along the canal pounds (the generic name for a level between two sets of locks) a set of paddle gearing is installed on the canal bank high above a stream into which the surplus water can be let off. The utility of some lets-off is well evidenced by their clean and well-greased state, others, long forgotten, will often have to be hunted for, through long grass and brambles. Fortunately, they are usually rusted solid, otherwise the results of malfeasance could be unhappy for people living in the valley below.

Drop planks were once provided at all suitable points along the canal bank for emergency use in the case of a breach or burst. In recent years it has been found that the local inhabitants are likely to convert them into fencing posts or firewood and it is now necessary to keep them at British Waterways Board depots.

A waste-weir would be placed above a stream to allow the

optimum canal level to be maintained, other than in times of flood, and it is these weirs which govern the depth of water in a canal. Some (control) weirs can be raised by movable boards to provide an extra head of water, either to feed another canal or to allow a flow to the local Water Board's reservoirs, for a good percentage of domestic drinking water, whilst starting its life in the hills, travels a canal before ending up in our taps. As the name indicates a stop lock served to hold the level of one canal higher than the other and was usually installed at the insistence of the first come to ensure that they did not lose any water to the upstart newcomer who was, in their eyes, a potential danger to their water supplies. This dog-in-a-manger attitude was to provide one of the greatest barriers to the free interchange of canal traffic and perhaps more than any other single cause was to lead to the success of railways. From the canal engineers' angle stop locks were nothing but a confounded nuisance. As an example, to reach its junction with the Trent & Mersey Canal the Macclesfield Canal has to cross over the Trent & Mersey by means of an aqueduct and then drop down via a stop lock (Hardingswood) to the lower level, an unnecessary and expensive flyover. The northern Stratford had to build one of the few examples of a canal guillotine lock, at Kings Norton, to placate the shareholders of the Worcester & Birmingham Canal. The gaunt remains of this stop lock look impressive against a sunset and are probably worthier of greater care than they get.

On the River Nene the problem is rather different and guillotine top gates are provided almost without exception to enable flood waters to be cleared quickly. Magnificently built, they are impressive if only for their utilitarian purpose.

Culverts are far from impressive, and more often than not resemble an enlarged rat-hole. Designed mainly to carry field drains or very small streams, the outermost courses of brickwork have probably long since been worn away and where subsidence is a problem the whole culvert may well be many feet below the surface of the field. Sometimes they can get completely lost, as at Glascote on the Birmingham & Fazeley Canal where a pound between the locks ran dry with monotonous regularity, the cause being believed to be a leaking culvert which had long since been built over by the motor-car factories on both sides of the canal. The majority of culverts carrying any quantity of water were built in brick and some, as shown in Telford's specifications, were made of oak. The modern culvert, perhaps carrying sewage or just a stream, will be in prefabricated concrete pipe. To insert this, unless a borer is used, it is necessary to stank off (close) half of the available waterway, dig it up, put in the pipe, rebuild the canal including the clay puddle, reopen that half and repeat the process in the other. The comments of the works foreman engaged upon the job have probably not materially altered over 200 years. Sometimes, too, things go wrong, as on the Grand Union Canal in 1971 when an emergency stoppage was announced because:

> During the construction of a tunnel sewer beneath the canal the bed of the waterway has fractured and it is, therefore, necessary to drain the canal between Castle Lane, Solihull (Bridge 83) and Woodcock Lane, Solihull (Bridge 86) from Thursday, 30th December and it is estimated that the works to the tunnel and the remedial work to the canal bed will be completed and the canal re-opened to traffic by Monday, 17th January 1972.[8]

When one of the last boatmen to use a horse-drawn craft on the Leeds & Liverpool Canal was asked what was the most important part of the canal, his answer was brief and abrupt: 'T'bloody towpath', and it is here that the change in waterways since their inception has been most marked. Some towpaths have over the years suffered from complaints that their engineers could never have envisaged. In the cuttings on the Shropshire Union Canal the perpetual running of water down the sides has eroded even the rock, in too many other places the towpaths have been broken away by 'shamanglers', but where the fishing from a canal bank is controlled by a responsible club the towpath is, more often than not, in a reasonable condition. The basic foundation of a normal towpath, which can be from 6 to 10 ft wide, will be broken rubble well pounded, behind a foot-thick layer of blue clay at the waterside. On top of this, according to the nature of the locality, we have small gravel or ashes.

When commercial craft were about it was the standard practice to tip the ashes from the range, morning and night, in any potholes that might arise – in the days of diesel engines this was probably an atavistic instinct left over from the time when a pothole could tear a horse's hoof, or worse, break his leg. Ashes also gave a good grip when one might otherwise slip and slide around a lock. Early canals had the towpath horizontal, but it was found that over a period of time they gradually gained a slope towards the canal and there have been many, many, times when the horse, far from pulling the boat, has been pulled into the water. Conversely, a towpath which is horizontal or sloping to the canal is better drained, and on the later canals where the towpath was cambered away from the water it became necessary to dig a ditch inside of this to carry away the top water, although the lessening of the strain on the horse more than justified this.

These then are the main factors which make up the specification of a canal, save one – and that is that the maximum flow of water on a canal, if it is to remain under control, must not exceed one half of one mile per hour.

2 *The Cost of a Canal*

The Caledonian Canal is the greatest undertaking of the sort attempted in the empire. It stretches S.W. and N.E. across the island from a point near Inverness to another near Fort William. It is chiefly formed by Loch Ness, Loch Oich and Loch Lochy. The total length of the canal, including the lakes, is $58\frac{3}{4}$ miles; but the excavated part is only $21\frac{1}{2}$ miles. At the summit it is $96\frac{1}{2}$ feet above the level of the Western Ocean.

It has been constructed upon a very grand scale, being 20 feet deep, 50 feet wide at bottom, and 122 at top; the locks are 20 feet deep, 172 long and 40 broad. Frigates of 32 guns and merchant ships of 1,000 tons burden may pass through it. This canal was opened in 1822. It was executed entirely at the expense of government, from the designs and under the superintendence of Thomas Telford, Esq., on whose skill and talents as an engineer it reflects the highest credit. The entire cost has been £986,924. It would, however, appear to have been projected without due consideration, and promises to be a very unprofitable speculation. During the year 1829, the total revenue of the canal, arising from tonnage dues and all other sources, amounted to only £2,575

6*s*. 4*d*. while the ordinary expenditure, during the same year, amount to £4,573 0*s*. 1½*d*! It is, therefore, very doubtful whether the revenue derived from it will ever be able to defray the expenses of keeping it in repair, without allowing any thing for interest or capital.

The following is a detailed account of the various items of expenditure on account of the Caledonian Canal, from 20th of October, 1803 to 1st May, 1830.

	£	*s*.	*d*.
Management and travelling expenses	36,691.	12.	10¼
Timber, and carriage thereof	72,317.	1.	10¼
Machinery, cast-iron works, tools and materials	128,886.	4.	7¾
Quarries and masonry	200,014.	4.	10¾
Shipping	11,719.	1.	6
Houses and other buildings	5,539.	10.	6
Labour and workmanship (day-work)	54,209.	1.	1¾
Labour and workmanship (measure-work)	418,551.	16.	8½
Purchase of land, and payments on account of damage	47,956.	12.	9¾
Purchase and hire of horses and provender	3,638.	12.	2¾
Incidental expenses	2,820.	18.	10
Roadmaking	4,579.	3.	6¾
Total cost	£986,924.	1.	6½[9]

This, then, was the cost of a 'Grand' Canal; the reason for its existence – to convey ships across Scotland; and a prophecy from a writer in the 1830s. The cost was accurate – note the recurring ¼*d*. and ¾*d*., a punctilious attention to detail that rings falsely in modern ears attuned to nothing less than 10p. As a ship canal it was outpaced in its building by the increasing size of vessels and the fruition of steam propulsion. As to the prophecy this was partly right, in so far as it never served for its designed purpose, and partly wrong for its value in wartime and to peacetime fishing craft should not be underestimated.

By contrast the costing of the Somerset Coal Canal, shown in the familiar 'orange-segment' fashion, is not final but does show the disproportionate sums that could be expended. Committee

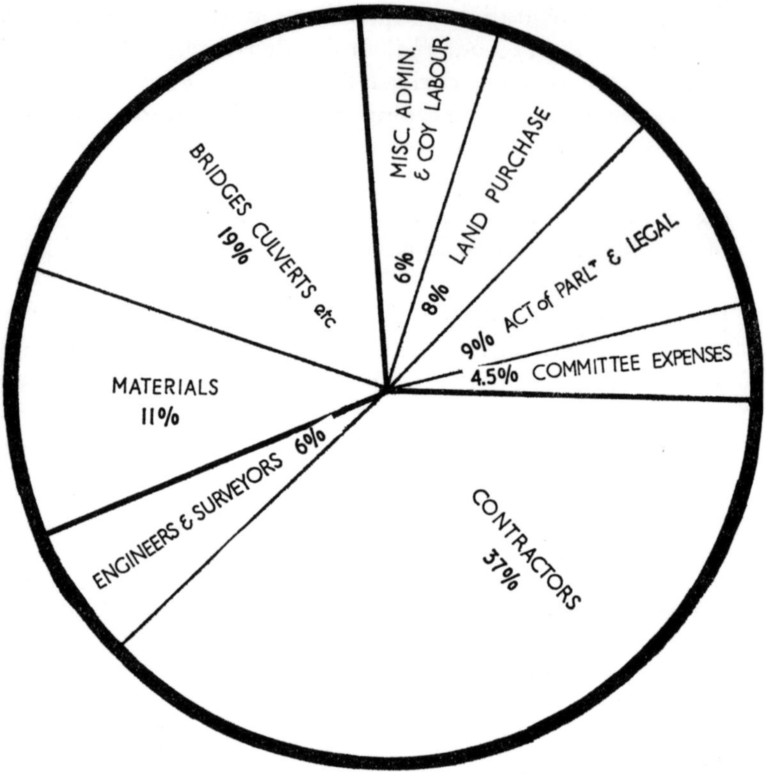

Fig. 2. The familiar orange segment.

expenses appear at first sight to be a trifle excessive. In fact this canal was playing with a mechanical lift invented by one Robert Weldon and specified in his patent No. 1892 of 1792 as a 'boat floated into trunk or casoon and lowered by means of rack and pinion'. It was not overly successful and the delays in completion, and hence traffic, led a vociferous minority of the directors to investigate the possibilities of either running a tramway in lieu of the lift (or even the canal) as propounded by Benjamin Outram, engineer of the Huddersfield Narrow Canal, or of an inclined plane. For the former they had to go to the North of England – a three-day journey costing, including hotels *en route*, something of the order of £32-odd, 'to view and investigate the works proceeding'. The shareholders begrudged this money, not surprisingly, for a labourer's weekly wage at this time was about eight shillings, and the 'Proprietors of the Navigation' were, in this case, mainly small men, rather than rich landowners, and doubtless they reasoned that the working of the lift should have been more clearly investigated in the beginning. In the event, in 1800, the lift was replaced by an inclined plane. Two years later the committee begged a further £45,000 to build a flight of locks in lieu of both lift and plane, remains of which are still extant at Coombe Hay.

It was not until 1910 that the main purpose of the committee's gallivantings was realised, when the GWR closed the canal and substituted a light railway. Other more reasonable committee expenses would include the hire of a public house for meetings – prior to obtaining their Act of Parliament – and the hire of an office or the purchase of same thereafter.

The first meeting of the co-proprietors of the canal from the Trent to the Mersey was held on 30 December 1765, a mere six months after Dr Irasmus Darwin, Josiah Wedgwood, John Wedgwood and Thomas Bentley had first employed James Brindley as surveyor. It was agreed at this meeting that an Act of Parliament for the work should be obtained and, incredibly quickly, by 14 May 1766 this was done. A month or so later we find Wedgwood cutting the first sod. Within a fortnight of the Act of Parliament a general assembly of the proprietors was held, James Brindley being appointed Surveyor General at £200 p.a., and Hugh Henshall Clerk of the Works at £150 p.a. It was decided that work must begin immediately on both sides of the proposed Harecastle tunnel and at Wilden Ferry, and furthermore an immediate call must be made on the proprietors for £6 per cent of their subscriptions. Work began on 17 July 1766.

This was an early waterway and, barring only the usual minor contretemps, it is unlikely that the committee, who were all businessmen, had a great deal to worry about. This was not the case where some waterways were concerned, an example being the Southampton & Salisbury Canal Navigation. Various plans for a canal line had at times been proposed and on 27 December 1792 a meeting was held at the Guild Hall, Southampton, where it was decided to seek five-guinea subscriptions from as many individuals as would give them to pay for surveys. These subscriptions would be counted as representing the first call on the shares which would eventually be issued. Various proposals were made, utilising Joseph Hill of Romsey as surveyor, who quoted a detailed figure of £47,208 17*s.* 10*d.* for the whole waterway; his estimate was accepted on 8 August 1793. A few months later he was called upon to prepare a small plan of the canal, from which an engraving was made and hand-coloured copies given to potential subscribers. This was the normal procedure and various such maps are still extant – for their time they are remarkably detailed.

By November 1794 Hill had revised his estimate and it had risen to £48,929 16*s.* 6*d.* This, in a simplified form, was:

	£	*s.*	*d.*
To digging and forming the Canal from the Town ditch at Southampton to the Junction in Houndwell and from thence to Northam	1,082.	0.	0.
To 3 Brick Archbridges in the above length will be wanted with a Stone coping and a Road made over them and Graveled	330.	0.	0.
To 3 Drawbridges for Occupation	190.	0.	0.
To 2 Locks one in the Town Ditch the other at Northam should be built with Stone front and backed with Bricks	1,700.	0.	0.
To digging the Canal from the Junction in Houndwell to the Mouth of the Tunnel	210.	0.	0.
To 880 Yards lineal measure in Perforating and making a Tunnel with a Nine Inches Brick Arch at £5. 5*s.* per yard	5,060.	0.	0.
To Digging, Banking, puddleing, lining and forming the Canal from the Tunnel at Southampton to the Andover Canal at Redbridge	2,350.	0.	0.
To an Aqueduct to be built with Stone at Shirleybrook	230.	0.	0.
To One Culvert at Millbrook Shore to be built with Stone	80.	0.	0.
To 3 Drawbridges for Occupation	190.	0.	0.
To 2 Brick Arch Bridges Coped with Stone and Graveled over them	220.	0.	0.

	£	s.	d.
To Digging, Banking, puddleing, lining and forming the Canal from the Andover Canal to Lockerly nearly in the direction as laid down in the Plan, this will be through various matters such as Peat, Gravel and Clay and Contains 90,136 Cubical yards at 3¼*d.*	1,220.	11.	10.
To Digging, Banking, puddleing, lining and forming the Canal from Lockerly to West Dean, this will chiefly consist of Gravel and Chalk in some deep cutting and will contain 64,120 Cubical yards at 3¾*d.*	1,001.	17.	6.
To Digging, Banking, puddleing and lining the Canal from West Dean to Alderbury Common which will contain about 130,100 Cubical yards at 3½*d.*	1,897.	5.	10.
To 317,100 Cubical yards of Deep digging in Alderbury Common, this I imagine will consist of sand Gravel and some part of a Terraqueous matter and will make no doubt a great deal of Water, this digging taking top and bottom at 6½*d.* per Cube Yard.	8,588.	2.	6.
To 318,162 Cubical yards of Digging, Banking, puddleing, lining and forming the Canal from Alderbury Common to Salisbury and which consists of chalk and Gravel at 3¼*d.* per Cube Yard	1,870.	18.	10.
To 146 feet of Lockage from the Andover Canal to Salisbury which I would propose to be divided in to 16 or 17 Locks as the Nature of the Land will admit of which will stand somewhat in the manner as represented on the Plan and Section . . . and will cost £67 per foot Vertical	9,782.	0.	0.
To 15 Arch Brick Bridges from Andover Canal to Salisbury	1,650.	0.	0.
To 16 Drawbridges for Occupation Roads	928.	0.	0.
To 6 Brick Culvers to convey the Water under the Canal from the high Lands	170.	0.	0.
To 8 Trunks to convey the Water to the Meadows and other places	160.	0.	0.
To One Aqueduct at the Test River, one at			

	£	*s.*	*d.*
Dunbridge River and one at the Laverstock River and a large Culver at the Test River to convey the Water to Lord Stawell's Meadows	1,323.	0.	0.
To Turning the River at Butt Green at the School Farm at Lockerly about 200 yards	150.	0.	0.
To Turning the Road at Dunbridge about 200 yards	80.	0.	0.
To 140 Acres of Land	4,480.	0.	0.
To two Reservoirs one at each end of the deep cutting at the Summit on Alderbury Common	620.	0.	0.
To Temporary Damages to Lands and Fences, Occupation Roads and unforeseen Accidents	3,366.	0.	0.
Extra cutting in Altering the line of the Canal at Redbridge	250.	0.	0.
	£48,929.	16.	6.[10]

Work then started and all, superficially, was peaceful. However, the committee should not have rested on their laurels, for being alarmed at various hold-ups concerning the building of the waterway they were forced to call in John Rennie, engineer of the Kennet & Avon Canal, to report on the whole of the works. This took place in March, in June they again called him in to have another look. By 1799 the whole of this waterway was in such a parlous condition that the company could not even raise the £21 necessary to pay for yet another surveyor, John Hollinsworth, who had been called in, to have a look at the works, and it is doubtful whether the final survey made by John Rennie in 1799 was ever paid for. The primary cause of the troubles of the Southampton & Salisbury Canal seems to have been Hill's bad first survey.

One of the biggest expenses which, in all bar a few odd cases, was absolutely inevitable, was the obtaining of the Act of Parliament authorising the works to proceed. However, this could, even allowing for all the triviata involved, be as low as £500, providing there was no opposition and no difficulties arose, but a typical cost of steering and preparing and obtaining an Act of

AN ACT

FOR THE

BETTER REGULATING THE NAVIGATION

OF THE

RIVER AVON.

WHEREAS the River *Avon* hath for many Years past been made navigable from the River *Severn*, through Part of the Counties of *Gloucester* and *Worcester*, to or beyond the Borough of *Stratford*, in the County of *Warwick*, for Boats, Barges, Lighters, and other Vessels, for the better supplying the said several Counties with Pit-coal, Iron, and other Goods and Commodities, for the public good of the said Counties, whereby the Trade in those Parts hath been very much inlarged and increased: **Preamble.**

And whereas frequent disputes have arisen between the Owners and Proprietors of the said Navigation, and the Persons navigating thereupon, touching the Rates of Tonage of Coal, and other Goods and Merchandize carried thereon:

For Remedy whereof, and for the ascertaining the Rates of such Tonage for the future, and to the end and intent the said River *Avon* may for ever hereafter be continued and preserved navigable, for the Encouragement of Trade, and the Benefit and Advantage of the Public in general; and that the Locks, Weirs, Bucks, Winches, Turnpikes, Dams, Floodgates, and other Engines, may for the future be kept in good and sufficient Order and Repair; and the Rents and Outgoings for Lands cut away, and other Rents, may be raised, and paid to the Persons intitled to receive the same;

May it please your MAJESTY,

That it may be Enacted; And be it Enacted by the KING's most Excellent MAJESTY, by and with the Advice and Consent of the Lords Spiritual and Temporal, and Commons, in this present Parliament assembled, and by the Authority

Fig. 3. Facsimile of first page of the Act to improve the River Avon Navigation, dated 1751.

Parliament where it was opposed was that of the Wilts & Berks Canal at £5,795.

In theory there were three ways of laying out a waterway. The first was simply to own the land on which it was proposed, no Act of Parliament would then be required and one merely employed one's own labour. The second was to obtain way-leaves, which involved renting the land through which the waterway was to run; such a route, unless you had the Parish Council in your pocket, must not cross any public highway. The third way was to obtain an Act of Parliament authorising the purchase of the land required for the waterway. Safeguards were built into these Acts, which protected the public right to cross the waterway and laid down the maximum charges that could be made for the traffic. Where there was opposition it was not uncommon for the dividends to be fixed at a given level, the surplus to be ploughed back into the waterway and/or to lead to the reduction of tolls. In practice there were ways around this latter. On the Glamorganshire Canal it was common practice to make the waterway toll-free for three months or so each year to keep the profits down. The proprietors of this navigation were, in the main, local ironmasters or mine-owners, who merely withheld sending their products down the waterway until the toll-free period, with resultant savings in costs to themselves. They adjusted this free period to suit their own requirements, and did not give prior notice of the dates involved.

Prior to obtaining the Act of Parliament it was needful for the Clerk, acting for the proprietors, to advertise the intention of the would-be shareholders to apply for such an Act. There were cases where the arrangement could be nobbled by opposition, the Clerk conveniently overlooking such a requirement, which could mean a year's delay before the matter could go forward, and sometimes simple negligence occurred. A summary of the bones of obtaining an Act is that the proprietors must be known by a certain name and style, have a common seal and have power to purchase lands, the name of the proposed waterway must be clearly specified, the number of shares defined and the price thereof. Incidentally, assuming one had a £100 share, even if one had already paid £94, should one fail to make payment of the last call of £6 the whole could well be forfeited.

On obtaining the Act powers would be given to enter upon and dig both the permanent and temporary works. This latter allowing for tramways to carry materials or lanes to give access to aqueducts and tunnels. A clause was often inserted binding the company to the line that was laid down in the plans, the deviation distance that was permitted being no more than 100 yds on either side. Surprisingly, although the width of the canal might be defined as being 10 yds, land could be bought up to 30 yds on either side. Wharves were also permitted to be laid

"I DON'T CARE HOW THIRSTY YER *@*!!*! COWS WERE — PUT IT BACK!"

Reproduced by kind permission of the "Farmer and Stockbreeder".

Fig. 4. Watering places for cattle.

down and quite often the charges that could be made for loading and unloading were defined. Some items had exemption from tolls – officers and soldiers, horses, arms, baggage and timber for use by His Majesty's Navy and government stores.

Mile-stones were a compulsory fitting in order that the tolls might be calculated accurately. Topsoil was protected, having to be returned either side of the waterway. Watering places for cattle could be called for. These, by the by, are anathema to the maintenance men of the British Waterways Board today, the erosive effect of the animals' hooves upon the bank being all too noticeable. Coalmining was often permitted to be carried on underneath the waterway, a rather short-sighted attitude as was found when the Two Lock Line on the Birmingham Canal Navigations disappeared in its entirety down a mineshaft. Steam or fire engines were permitted, if required, to throw the water back up to the top, or any intermediate length. Sewers had to be defined while the right to withdraw water from streams and rivers was often restricted. To protect the proprietors it was laid down that persons tampering with the works were liable to be transported for seven years.

Conveniently for later researchers, two copies of the proposed plan of the waterway and the book of reference listing all the landowners along the route had to be deposited with the Speaker of the House of Commons.

Quite often there was strong opposition – we have already seen that Telford was forced at great cost to the proprietors to divert the line of the Birmingham & Liverpool Junction Canal, and certain mill-owners virtually brought the downfall of the first canalisation of the River Stroudwater by refusing to allow locks to be built, although they were later to take a more reasonable view on the matter; a further Act had to be obtained by the proprietors which still included clauses protecting the millers. One of these reads:

> . . . provided always, That the Owner or Owners, Occupier or Occupiers, of any such Mill or Mills, Shall and may . . . lawfully demand and receive from the Said Company, as a Recompence and Compensation for such drawing up, opening, putting down, or shutting, such Clough or Cloughs, or Floodgates respectively, after the Rate of Two Shillings an Hour for the First Twelve Hours, and after the Rate of Three Shillings an Hour for every Hour afterwards, that the Mill or Mills on each Pen or Pound of Water shall be thereby stopped or hindered from working . . .[11]

Sometimes an Act would permit the waterway to also run 'Rail Ways or Stone Roads', as on the Brecon & Abergavenny

6 Success story. Neath Aqueduct on the Tennant Canal, *c.* 1811

7 An attempt that failed: Murtry Aqueduct on the never-opened Dorset & Somerset Canal

8 Tramroad and canal at Penydarran, Glamorganshire Canal

9 Three-in-one bridge. The canal (Forth & Clyde), the railway and th river (Luggie Water) at Kirkintilloch

10 Bollard

11 Water gauge

Holt Lock, River Severn

12 Traffic lights for craft entering the lock

13 Remote-controlled paddle-gearing

14 Plan for a lift utilising boats, by Robert Fulton, 1796

Canal. It is also of interest to note that quite often the engineer would himself buy shares, as did Thomas Dadford in that canal.

The fencing of towpaths was most strictly defined and should they not be made the landowners neighbouring the waterway had the power, after a space of three months, to put up such 'Fences, Gates, Styles, Bridges, Arches, Passages, Watering Places and Other Conveniences' as they wished at the company's expense.

Power was given in most Acts for the committee to borrow money, the total sum being clearly defined. One canal that received two Acts and spent quite a lot of money without getting anywhere at all was the London & Cambridge Junction Canal, which was to have run from Bishops Stortford to Cambridge. The first survey was carried out by Mr Whitworth under the orders of the Common Council of the City of London and a later survey was carried out by Robert Netlam and Francis Giles on behalf of some private citizens. The waterway was empowered to raise £575,000 and if this was to prove insufficient they could borrow a further £300,000 without obtaining a further Act, but, and again it was a very big but, no works were to proceed until £425,250 had been actually subscribed. It was not.

A waterway which was easy and successful in its work might only require one Act of Parliament: the Erewash Canal, for example, received its Assent on 30 April 1777, and was quietly dug and worked thereafter. By contrast the Oxford Canal works required no less than eight Acts of Parliament between 1769 and 1829.

A very small expense to a waterway was that of providing stationery and printing. A number of advertisements were necessary, mainly owing to the relatively inefficient postal service of that time. It will be understood that at the first meeting of the proposed proprietors of a navigation, the amount of the shares would be decided, together with their cost, which we will take as being £100. At the meeting the first call of £5 or £6 would be made, and further calls made from time to time as the works proceeded, utilising advertisement. Some would be curt in their manner – the Trent & Mersey advertised in *Aris's Birmingham Gazette* in October 1769:

Navigation from the TRENT to the MERSEY.

In pursuance of an Order of the Committee of the Company of Proprietors of this Navigation, bearing Date the First Day of this Instant November, All and every the said Proprietors are required to pay into the Hands of Mr. Josiah Wedgwood, of Burslem, in the County of Stafford, their Treasurer, Francis Cobb, Esq. in Lichfield; or Messrs. Ashton, Hodgson, and Co. in Coleman-street, London, the Sum of Six Pounds per Cent. of their respective Subscriptions, on the eighth Day of January next; and the further Sum of Four per Cent. on the second Day of February next.

JOHN SPARROW, *Clerk to the said Company.*

Some would be frightening, in effect 'If you do not pay, all you have spent is lost'. Sometimes the Secretary's Report would be rather sad, for instance on the previously mentioned Southampton & Salisbury Canal, when they were desperately scratching around for money, most of the proprietors, preferring not to sink good money after bad, either ignored notices and letters, or told the committee what they could do. The Secretary was forced to report, in 1799, that proprietors had written varied answers to letters, including 'I can only say I believe we have fallen into bad Hands and that I have no money at present', 'I decline risking any more money in the undertaking of the Southampton & Salisbury Canal' and that only one man, who was totally ignorant of everything relating to the canal, was willing to risk paying up.

Canal-cutting is the first expense that is thought of, although it rarely accounted for more than one-third of the total cost of any one waterway. Various treatments of this matter were possible, some of which depended on the length of the waterway and others upon its nature. When the River Rother was being canalised, Lord Egremont, who owned a large percentage of the land through which it passed, used his own labour force, which he already had to hand for the maintenance of his estates. Its entire length was 11 miles, which made it a radically different problem to that of the Trent & Mersey Canal at 93 miles. A second approach, again more feasible on short lengths, was to allow the engineer to be the main contractor, he being responsible for the hire of labour, supply of equipment, etc. This had advantages to the committee inasmuch as they had only one man to deal with, but it had the disadvantage that if the engineer were a little slaphappy, or was often called away on other works, or was just downright incompetent, they had no real way of checking. The third method was to let the whole work out to one large outside contractor; this was the method favoured later in the railway age. The fourth, and most commonly used, method was to obtain tenders for a given section of the waterway, accept that which in the opinion of the engineer seemed the most satisfactory (not always the lowest, of course) and leave it to the engineer or his assistant to ensure that the work was well done.

The contract, which, using round figures we will say was for £1,000 per mile, would stipulate some matters being in favour of the committee and others in favour of the contractor. It could be that the committee were to find the materials, the contractors finding the men, or, alternatively, the contractor might be responsible for both, subject to the final say of the engineer. The normal system of payment, irrespective of which method was used, was that 10 per cent would be paid in advance and further sums thereafter, as the works proceeded. Of the whole contracted price 10 per cent was withheld for a period of twelve months to ascertain whether or not the works had settled satisfactorily. Usually a built-in float was allowed, whereby the contractor could increase his estimated cost by a given percentage, sometimes as low as 5 per cent and on other occasions as high as 33⅓ per cent. If, in the opinion of the engineer, the works were not proceeding as they should, the company invariably had the right to take the contractor off the work, continuing with their own, direct labour or using a replacement contractor. The right of the main contractor to sub-contract was usually limited. Clauses allowed withdrawal from the contract in the event of death of a principal within the contracting firm or if the line of the canal were not in accord with that stated by the engineer, in which case the contractors were recompensed for the work done.

Some contracts which were placed tended to be rather vague, reflecting either the haste with which the work was commenced or the engineer's attitude. Others would go to the point of defining the type of brick to be used in locks, or the number of pieces of lead into which a gate anchor was to be embedded, or might stipulate that only Derbyshire grit-stone was to be used for the lock chambers.

Where the contractors themselves were concerned, it must be remembered that these men were not like McAlpines or Mowlem, the great motorway contractors of today. Usually they were men who had come up the hard way, perhaps being responsible for the resurfacing and maintenance of part of a turnpike, or the building of the local brick- and lime-houses. Some were later to become relatively successful, and the engineer might take them from job to job, knowing that their workmanship was good. Others would dig a few miles of waterway and than fall back into obscurity. Too many went bankrupt.

The price of bread was, to the working man, the governing factor in the cost of living. Its average rate is shown below (all figures are in sterling, per 4 lb. loaf).

	d.
1760–64	4·9
1770–74	6·8
1780–84	6·7
1790–94	6·6
1800–04	11·7
1810–14	14·6
1820–24	10·0

These figures are based on prices prevalent in London and, indeed, prices varied over the countryside, for on 10 November 1818 we find that in Northampton wheat cost 86*s.* 2*d.* per quarter, in Berkshire 83*s.* 6*d.*, in Oxford 81*s.*, Wiltshire 75*s.*, and in Essex 73*s.* 10*d.*[12] The local differences quite often can explain some of the short-haul traffic that occurred on waterways, the Wilts & Berks Canal clearly profiting by the movement of wheat between the two counties. The cost of oats could also have a material effect upon the profit or loss of the contractor. It is estimated that a normal contractor had two horses to every man employed and these animals had to be fed. It is impracticable to give all the economic factors that were involved in the price of works upon a canal, minor items as well as the major could each take their toll. Navvies needed boots, boots had to be made, as the price of wheat increased the miller increased his charge and the flour cost more, therefore the labour to make the boots cost more, higher wages being paid because of the higher prices. The navvy who wore the boots then wanted more money to pay the higher prices for flour and for boots, this later item showing a fluctuation in price from 4*s.* in 1765, 3*s.* 6*d.* in 1785, 4*s.* in 1795 to a peak of 5*s.* 9*d.* in 1805. It was not only the navvy who wore boots and the men in the brickyards who wanted more money – the men at the ironworks did also. The average lock contains about one ton of iron – paddle gearing, paddle rods, gate anchors, hand-rails, straps, bolts and other bits and pieces. The iron-master smelting in Wales would find his costs had risen and so all the iron-work would cost more. To make it worse, due to the profit structure of most companies, prices rose out of proportion to the basic effect of putting the boots on the feet of the labourer. Assuming that mining the iron-ore cost £1, the labour to convert it to paddle gearing a further £1 and that depreciation, factory overheads, etc., accounted for another £1, the profit would then be based on a total of £3. The iron-master would probably consider £1 profit to be reasonable, making a total selling price of £4. Given an all-round increase in the cost of living of 10 per cent, mining the ore would then cost £1·10, the labour and overheads, etc., the same amount each, giving a total of £3·30. The iron-master still requires his 33⅓ per cent profit, raising the total cost to £4·40. Thus from small acorns great oaks grow and 10p at the pit-head becomes 40p at the factory door. Allow for proportionate increases in the cost of transportation and contractors' labour, and it will be understood that the poor contractor working on, probably, a 5 per cent overall profit, quickly has his profit turned to a loss. With such inflationary rises in costs the simplest and most obvious way for a contractor to ensure his profit was to skimp the job. However, if the engineer was a reasonable man, he would go back to the proprietors and say: 'This is rather hard luck, the contractors are doing the best they can, everything is going up, will you allow me to increase the contract price by 10 per cent?' He would probably point out that, in turn, the proprietors would gain more revenue by increased tolls. He might also explain to them that to take the contractor off the job at that moment would involve them in more costs, having to employ their own men, find the materials, etc., than would be involved in paying the extra 10 per cent. Telford was quite pugnacious in this approach and he could afford to be so, as he had the confidence of the proprietors and he knew his contractors. Other engineers, Dodd and Giles – are two – would make rather feeble excuses as to why the work was not being completed. We have mentioned before concerning the Southampton & Salisbury Canal that the shareholders were growing a little suspicious in 1798, and Rennie then reported, apropos of the tunnel at West Marlands that:

> In respect to the work already done, it is by no means complete, those parts that are likely to stand are ill framed and seem to have been done with little care or judgement . . . In joining the different lengths of Arching together they do not in many places agree, i.e. sometimes one length is sunk more than another . . . At the West end of the tunnel a part of the sheeting for about 16 or 17 yards in length has entirely risen up and the tunnel has sunk about a foot. The whole of this length must be taken out and done anew . . . The sand that has been used for the mortar is perfectly unfit for the purpose, being little better than clay . . . An agent or superintendent skilled in works of this sort should be procured, with an adequate salary, and his whole attention directed to it, not only to see the works are properly executed, but that no improper materials be used . . .[10]

The engineer, Joseph Hill, was hauled over the coals and he could do little but pass it on to the contractor, Thomas Jenkins, who was told bluntly that he had to put the defective work right at his own cost and that no further payments would be made to him until he had done so. He, in turn, found he could not pay his men or pay for materials and he went bankrupt. Perhaps rather sadly Joseph Hill was two years later a debtor in the Kings Bench prison.

Other contractors found something of the same problem. During the building of the Leicester & Northants Union Canal riots took place in 1793 and 1795, mainly on account of the poor wages being paid, this being an inept endeavour by the contractor to keep rising costs down. On the Wilts & Berks Canal, Robert Whitworth, engineer and also contractor, stated on 6 June 1800 that:

> The canal is now open to the turnpike at Dauntsey except for 500 yards from England's land to Chippenham . . . and about 200 yards of the line to the turnpike road in the town of Calne near a mill called Calne Mill. The only work remaining to open Bowd's Farm is the completion of the lock and bridge at Dauntsey, to be done in a few weeks. The canal is cut from Trow Lane to Tockenham and a considerable amount of earth is to be carried in boats, which is most desirable because it is the cheapest mode, and by that time the canal may be made navigable to the turnpike road near Hunt's Mill. Ten boats are on hand. The total expenditure to date is £55,856, this exceeds the estimated price due to the fact that the brick tax has been raised, the price of labour has risen by 20%, and that the diversion of the Calne branch and the tunnel introduced an extra expense of £1,800. Also all coal and stone had to be carried partly overland, but the Kennet & Avon and the Coal Canal expected to be open in three months reducing these transport costs. No difficulty is expected from Hunt's Mill to Abingdon which may be completed in less than two years.[13]

Even here though the engineer could make mistakes, as we find in a letter from James Johnson, dated 27 January 1796, when he found that workmen had entered his premises and begun to dig for clay. He was a trifle peeved as the company had not even agreed the price of the land, let alone paid any money. When this particular waterway finally opened in 1810 it had cost £255,263 rather than the £111,713 originally quoted by Whitworth in 1794.

Fundamentally, it is true to say that all the early canals were dug by the muscle power of men and horses but this is not to say that machines were unavailable, for we find in the *Monthly Magazine*, vol. ii, the following account of the operation of E. Haskew's patent excavator:

> This machine takes the soil from the bottom of the canal, at 40 feet deep, with equal facility as at six feet from the surface! One of them is at work upon the Gloucester & Berkeley Canal. By the assistance of two men only, it removes 1400 loaded barrows from the bottom of the canal, to the distance of 40 feet, in 12 hours; and is so contrived, as to take up the loaded barrows, leave them at top, bring down the empty ones in regular rotation, and leave them at the bottom. It can be moved along the canal to the distance of 26 yards in 10 minutes, by the two men that work it.

In October 1793 Mr Joseph Sparrow took out a patent for a machine, consisting of a box with its bottom on hinges, suspended by a sort of universal gib or crane for elevating and discharging the soil dug out of the canal. While in 1796 Robert Fulton, that somewhat eccentric if ingenious engineer, was writing that he had a machine planned which could dig canals and quoted a couple of good reasons for its use. 'But if Such a machine or Any other Can be made to Answer, it will Give further advantages by expedition and Set one independent in a great measure of Canal diggers. But in America where Manual Labour is of So much Importance Something to Assist in Cutting would be a most material improvement.' In Britain, although the cost of manpower had risen, there was never any shortage of men and they continued to do the work the way it had always been done.

The first canal engineers avoided cuttings and embankments as much as possible, this being painfully apparent on such waterways as the Southern end of the Oxford and the Northern end of the Stratford-on-Avon, where the waterway will meander around in a great arc to avoid a hillock probably not more than 10 ft high. Neither was there any great attempt to use any soil excavated to make embankments. Assuming that one had a strict engineer the contractor would first remove top soil – as he was obliged to do – and place it somewhere for safe-keeping.

Digging would then commence, the workmen employed on the job being known as navvies or navigators. The navigators usually worked in a gang, having sub-contracted, or undertaken, to shift a given amount of soil within a certain period.

> We now approached the great tunnel, which forms part of the communication, between the Severn and the Thames . . . one end penetrates the hill at the village of Saperton, the other end comes out in Heywood; we turned on our left to visit the former, and saw the shafts busy in several places, at a distance of about 230 yards from each other; by this means they wind up the materials from the cavity and expedite the work. The earth is principally a hard blue marle, and in some places quite a rock which they blow up with gunpowder; the depth of these pits are upon an average eighty yards from the surface. The first contractor received £7 per yard from the company, and the labourers rent it at the rate of about £5 per yard, finding candles, gunpowder, &c., the workers are in eight gangs, having two or three reliefs, and continue eight hours at a time, day and night.[14]

On later canals where deep cuttings were involved, barrow runs would be utilised, simply a plank laid by the side of the excavation, with a horse-gin, or jenny, at the top. A line would be run from the horse, either via the gin or over an ordinary pulley to a barrow at the bottom of the plank. A navvy would fill this barrow, the horse would move forward at the appropriate command, drawing the barrow up the plank, the navvy guiding it as it went. If horse, barrow or man slipped an accident was almost inevitable. Again, in a cutting, it was the habit of the navvies, in order to save time, to undercut a bank, and not a few were suffocated when the overhang slipped on to them. Blasting through rock was carried out in an incredibly primitive manner; a hole was knocked into the rock and gunpowder and fuse were pummelled in with a copper or iron rod. The fuse was then lit and the resultant explosion was supposed to loosen the rock. Maiming and death were, of course, attendant risks, for the fuse might go off prematurely, the gunpowder object to the ramming process, the rock not explode in the manner expected, or too short a fuse be used to save time. A casualty rate of one of two deaths per mile was considered reasonable during excavating and passed more or less unnoticed.

I've navvied here in Scotland, I've navvied in the south,
Without a drink to cheer me or a crust to cross me mouth,
I fed when I was workin' and starved when out on tramp,
And the stone has been me pillow and the moon above me lamp.

I have drunk me share and over when I was flush with tin,
For the drouth without was nothin' to the drouth that burned within,
And whene'er I've filled me billy and whene'er I've drained me can,
I've done it like a navvy, a bold navvy man.

I've met a lot of women and I liked them all a spell –
They can drive some men to drinkin' and also some to hell,
But I never met her yet, the woman cute who can
Learn a trick to old Nick or the bold navvy man.

I do not care for ladies grand who are of high degree,
A winsome wench and willin', she is just the one for me,
Drink and love are classed as sins, as mortal sins by some,
I'll drink and drink whene'er I can, the drouth is sure to come –

And I will love till lusty life runs out its mortal span,
The end of which is in the ditch for many a navvy man,
The bold navvy man,
The bold navvy man,
Safe in a ditch with heels cocked up, so dies the navvy man.[15]

Some engineers instructed their contractors that the navvies must always be paid in cash, others turned a blind eye to the Tommy Shops. The principle of the Tommy Shop was excellent; goods could be bought in bulk – flour, bullocks, potatoes and beer – and resold to the men at cost, rather akin to shopping in a Cash-and-Carry store today. The practice, more often than not, was very different. The contractor, to avoid having to pay out too much money, might quite possibly pay the men half in cash and the other half in tokens redeemable at the Tommy Shop. He would buy poor quality goods, water the beer and expect to make 100 per cent profit, thus often subsidising the losses on his contract with the profits from his Tommy Shop. The system should have been advantageous to the navvy, for if he started work on the Monday with no money he could go to the Tommy Shop and be given goods on credit, the sum being deducted from his wages; but given an unscrupulous contractor, the navvies' inability to read and the fact that, in all probability, he was the worse the wear from drink, the reckoning at the end of the fortnight or month – on 'Pay Saturday' – might well leave him with nothing in hand, and so perforce the credit bill would run up again.

You boatsmen and colliers all,
Come listen to my ditty,
I'll sing you a song before its long,
It is both new and pretty;
It is concerning Tommy Shops,
And the High Field ruffian,
He pays you with a tommy note,
You must have that or nothing.

With the colliers I begin,
How they pay each other,
Nothing have they but a tommy note,
From one week to the other,
On Saturday when a week's work is done,
And to receive their money,
The High Field devil has learned a trick,
To pay them off with tommy.

The boatsmen now I bring in,
That sails from High Fields to Runcan;
The boatsmen and their wives,
They curse him at the junction,
And all belonging to the branch,
That know the art of boating,
Wishing the tiller down his throat,
It would be a means to choak him.

When they had done their Runcan voyage,
And go to receive their money,
One half stops for hay and corn,
The other half for tommy,
Then to the tommy shops we go,
To fetch our week's provision,
Their oatmeal, sugar, salt and soap,
Short weight and little measure.

Saying if we had money instead of this,
Provision we could have plenty,
The profit they get out of us,
Is nine shillings out of twenty,
Then we jump on board the boat,
And the children look so funny,
The voyage we so cheerful go,
Till we have eat all our tommy.

There is one amongst the rest,
That knows the art of boating,
He vows and swears a wife he'll have,
So long he has gone a courting,
He vows he will married be,
Come listen to my joke sir,
And when the parson's done his work,
I will pay him with a tommy note sir.

Now we have finished our voyage,
The children look so funny,
For here at Runcan we do lie,
And have eat all our tommy,
Come gear the horse and clear the line,
And jump on board the boat sir,
Both night and day we'll stear our way,
For another tommy note sir.[16]

As we mentioned before, not all the work would necessarily be put out to contractors. Indeed, it might be impossible to get them to quote. Again, sometimes they would have to abandon a contract and such works would be finished by the company. Assuming that the company was far-seeing, it would take on a number of workmen whom it would plan to utilise in the future; thus the company might insist that all the lock gates and stop gates be made at its own workshops, that its own bricklayers be responsible for the bridges and masonry of lock walls. A change of engineer midway through the building of a waterway could lead to many difficulties – the contrast between the northern and southern halves of the Stratford-on-Avon Canal makes it difficult to realise that they were once under the same ownership. The locks and bridges differ radically from one end of the Trent & Mersey Canal to the other, partially due to differing local materials being available, but also to the habit of this company of sub-contracting whole sections of the work to each contractor to do as he wished within the permitted tolerances of the engineer.

Railroads of varying types were an essential adjunct to any waterway during its building for the movement of spoil from cuttings or to embankments and for the movement of materials from the quarry to the line of the canal. In the case of a number of waterways it was found advantageous to substitute rail or tramways for part of their intended canal line, the Ashby-de-la-Zouch Canal being one. 'When authority was first obtained, for the making of this canal, it was the intention of the company to have continued the canal to the places mentioned in the title of the Act, with 252 ft of lockage. They, however, adopted railways for all the branches where lockage was necessary.'[4] In this case it was partially to save money but equally because they apprehended difficulties in the supply of water. In the case of the Somerset Coal Canal one feels a certain degree of desperation; after their failure with lifts and inclined planes they were faced with having to build a flight of locks to complete the branch to Radstock. Initially they laid a railway to connect the portion of the branch already built to the main line, subsequently substituting a railway for the whole distance. During the delays in the building of Blisworth tunnel on the main line of the Grand Junction Canal a railway was laid over the hill, goods being transhipped at both ends, but the complaints received by the proprietors were such that work on the tunnel had to be speeded up with extra shafts being dug to enable the work to proceed from more faces.

The cost of the land which was required for a waterway could vary to a very great degree. A quarter of an acre might cost £500 if the owner knew he, or she, could get it; while other landowners were pleased to give the land to the canal company, sometimes being reassured of the benefit to themselves – especially if they were one of the proprietors – or receiving an indirect payment by the canal being 'beautified' where it passed through their land. There is a (possibly apocryphal) tale of a clergyman through whose land it was proposed to cut a canal. He insisted that each of his fields which was to be divided by the canal should be provided with an accommodation bridge to give access from side to side. As the waterway slowly ground towards his land he discussed the first bridge and decided it was not altogether necessary and proposed to the engineer that if the company would pay him half the cost of the bridge he would be prepared to forgo such a convenience. Oddly, but not unnaturally, one by one, the five bridges were found to be unnecessary provided the same terms were met!

It must be mentioned that some landowners – particularly Quakers – took a very great interest in the navvies who were at work, arranging for a surgeon to call once or twice weekly or paying some small sum to the local cottage hospital where the more battered navvies were sent. In Shugborough House in the Billiard Room can be seen the life-saving set that was issued at the time of the building of the canal; whether it could, or did, save life is a moot point but the thought was there.

'Puddle inspectors and counters of workmen' were two groups of men most vital to the success of any waterway. In theory they existed independently of both engineers and contractors, although more often than not they were appointed at the recommendation of the engineer. The clay puddle in the bottom of the canal is its life and death, whilst counters of workmen were there to ensure that the contractors kept a good number of men at work. Being paid by results the contractors were prone, when they could, to put all their men to work in one section to complete that before paying any attention to the balance. This did not always please members of the committee as they liked to see lots of work going on all at once. That 'penny-packets' were useless on their own was irrelevant – it looked better!

Sometimes indeed, a contractor, knowing that an inspection was going to take place, would divert a gang of navvies from the main site to a nice easy spot somewhere, where they could make a big hole quickly; as soon as the inspection was over the men would return to the main work.

Clerks were a necessity, for all letters had to be handwritten, files had to be kept, cheques made out and, most important of all, cash kept in hand, or at least got ready, for pay days. It must be borne in mind that there were very few large banks at this time and a small bank did not keep a large amount of cash on the premises, sending instead to the nearest county town when it was required. If direct labour were employed, 2,000 men might be demanding payment at the same time, and it is on record that the Dragoons were called out to protect the Clerk when it came to pay day. Equally, sometimes the Dragoons were needed when the contractor had to explain to his navvies that there just wasn't any money – either he had gone bankrupt or the money had not arrived from the bank.

These are, of course, only a few items involved in the cost of a canal. Boats were often provided by the company at a cost of about £30 each, stabling could swallow up quite large sums and the party inevitably given at the opening of a canal did not come for free.

> The opening of the Wilts & Berks Canal into the Thames at Abingdon was celebrated there on Friday with demonstrations of joy suitable to the completion of so important a part of inland navigation. At half past two o'clock a body of the Proprietors passed the last lock into the Thames amidst the loud huzzas of multitudes assembled to witness the spectacle. The party then left the canal and proceeded to the Council Chamber where they were joined by a numerous assemblage of gentlemen, including members of Parliament for Cricklade, Abingdon, Oxford, Hereford, Ludgershall, . . . and partook of a splendid dinner prepared for the occasion . . . The day was spent with great conviviality and harmony, enlivened by many appropriate toasts and songs, and the Company, high gratified, separated at a later hour.[17]

3 *Fixed Assets of a Canal*

> It is with pleasure we congratulate the Public on the Probability of Coal being brought by Water, near this Town, in a few Days; and that the Canal Company have not only resolved to sell the same this Winter at their Wharf for Fourpence Half-penny per Hundred, long Weight, of 120 lb. but to fix the Price of their Delivery in every Street thereof: And in order for the better accommodating of the Poor they have determined to establish Coal-Yards in different Parts of the Town as soon as possible, where it will be sold in Quantities so small as Half Hundreds or less: And, indeed, there is great Reason to believe, that the Price of Coal will come (after the present Winter) cheaper than Four-pence Halfpenny per Hundred; and that the Gentlemen who have the conducting of this important Affair, will use all possible Means to prevent impositions of every Kind.[18]

Having got over their excitement at this lowering of prices, no doubt the populace retired to a near-by public house to celebrate, and where better than that situated on a near-by canal – the Eight Locks at Oldbury, the Eagle and Sun on the Worcester & Birmingham Canal, the Junction Inn on the

15 Lifting bridge at Wrenbury on the Shropshire Union Canal, Welsh Section

16 West-bound fishing boat leaving Fort Augustus Top Lock, Caledonian Canal

17 Vandalised boatmen's lodging house at Atherstone, Coventry Canal

18 Garstang, Leicester Canal, *c.* 1900

19 Warehouses at Shipley, Leeds & Liverpool Canal

20 Torksey Lock at the entrance to the Fossdyke Canal

H 154

Shroppie, the Boatman's Rest on the Stratford Canal, or one of the innumerable Navigations. But a more regular user of these public houses was the boatman, especially in horse-drawn days. Trips usually started at the company wharf and a day's journey was governed by the location of suitable stabling.

Fodder represented much less of a problem, for an allowance in gauging the tonnage of a boat was always made for its carriage and sufficient for a trip would be carried. Some – a very few – steerers who were on a regular run and whose family was excessive for the cabin accommodation used the pubs to stay in overnight. A number of very remote pubs not only came into being with the canal, but were dependent upon it for their living. As the number of boats declined they closed. On the Whitton (Long Buckby) Flight on the Grand Union Canal there were seven pubs in the 1920s, the last of which closed its doors in 1970; the short season of pleasure-craft cruising was insufficient to keep it going. Others, proper boatmen's pubs, which have disappeared recently include the Bull & Butcher at Napton (Oxford Canal) and the Halfway House at Tardebigge (Worcester & Birmingham Canal), leaving this latter canal without a pub for three miles and 36 thirst-making locks! Another hostelry, on the Shropshire Union Canal, even these days finds it necessary to state 'This Public House is open Friday, Saturday and Sunday 11 a.m.–2 p.m. and 6 p.m.–11 p.m.' – the publican having other work during the week.

Some, moving with the times, would be unrecognisable to the old-timer, shorn of stables and with yards converted to car-parks. Any connection with the boats and boatmen is tenuous, typified by a collection of 1970 Birmingham-made martingales and 'pretty' boat paintings. On the broad canals and rivers of the north a happier state is preserved and one is certain of, in season, a good fire and cold beer.

The importance of public houses as warehouses on a canal, serving the local populace, is often overlooked. Surely one of the most unusual was on the Wilts & Berks Canal at Wootten Bassett where the cellars of the Bridge Inn were used for the storage of spices. Presumably these must have been for the use of the gentry, but we are told that in 1821 the working class were most distressed through the non-arrival of salt. The

21 Stoppage at Braunston, Grand Junction Canal, *c.* 1920

waterway was suffering, as was its way, from the drought and nothing was moving.

Some public houses, although dependent upon waterways for their existence are, in contradiction to the old Navigation, Wharf or Three Locks, named after outside interests. The Greyhound, the Crown, the Globe and the Praed Arms are all on the canal-side, although the latter was named after the first chairman of the Grand Junction Canal Company.

The vicissitudes of a canal-side hostelry have been many and varied. Some have lost their licences through declining trade, others following the death of the owner and many have been eliminated by the brewers who have transferred the licence to a modern plastic and chrome palace on the near-by road. The Navigation at the Old Wharf, Tardebigge, on the Worcester & Birmingham was built when the canal terminated at this point, and in its latter days it provided a convenient habitat for the 'leggers' waiting to propel the boats through Tardebigge tunnel. Unfortunately, there were two drawbacks with the usage of this waiting place, the first being that the leggers became intoxicated and some drowned, the second, the logical corollary, was that being intoxicated they enjoyed a good fight and when they extended their fighting to include the natives of the district the lord of the manor, Lord Windsor, who owned the land whereon the public house stood, closed it, removed its licence and demolished the building. Normally leggers had small brick-built huts, situated at one or both ends of a tunnel, wherein to wait, together with the animals, whether mules or horses, that had been brought over the tunnel. Leggers were official employees of the Canal Company rather than of the carriers or the boatmen and were paid so much per boat, the sum varying according to the distance and time involved. On the majority of canals they were issued with large brass amulets and worked on a rota or shift system, some form of organisation being necessary as the situation could easily arise where all the leggers were on one side of the tunnel. If a boatman was working his boat single-handed he would pay a legger to take the animal across the tunnel, but it was more normal to carry a lad who would undertake this work. George Bate has gleefully told us of one serio-comic adventure that befell him:

On this particular day my father steered the boat through the two tunnels, while I took the two donkeys over the tunnels. All went well until I got on the top of Shortwood tunnel, there are two clap gates that close the fields from the towing-path, I got through one gate with the donkeys alright, but on getting to the second gate I had to go past alongside the donkeys to open it, and this was when the trouble started. Those donkeys started off at the gallop towards Stoney Hill with me after them. I never got hold of them, still and docile again, until they had been running as far as the top of Scarfields Hill, Alvechurch. When I finally got to the other end of Shortwood with the donkeys my father was still waiting with the boat – after two hours. Of course, I had to do some explaining and tempers were not at the best.

George Bate is still alive and lives on the Worcester & Birmingham Canal in a cottage connected with the reservoir, known as Reservoir House. It is situated by the side of the reservoir and one of Mr Bate's duties was, and still is, feeding water to the reservoir in times of surplus or from the reservoir to supplement the flow going down the locks. His cottage is unusual in so far as it pre-dated the building of the waterway, being purchased from the landowner after the completion of the canal. Another reservoir house in the same vicinity is that at Bittell; it is unusual in being built on an island. To the side there is Jacob's Cut, a short branch leading to the old pumping house, in front is the main line of the Worcester & Birmingham Canal and to the back and the remaining side the Bittell Reservoirs, only one of which has ever served to feed the canal, the other being built to recompense the mill-owner for the withdrawal of water from the local river. Ironically it is only within the last decade that running water has been installed within the house, supplies previously having been fetched in a tank mounted on a cart from upwards of two miles away. This lack of water supply to canal-side properties is not uncommon and was probably of little importance in the days when working boats regularly went up and down and there were local springs to be drawn upon to supplement the well supplies; nevertheless it is of small wonder that lock-keepers preferred the neighbouring hostelry. However, before the days of industrial and farming pollution, cut-water was quite adequate for normal washing and domestic purposes. The lack of toilet facilities is, without running water, obvious, and is a matter which has not entirely changed – not over-surprisingly when it is recollected that maintenance men going about their normal duties on the canal are expected, at the best of times, to use the hedge for such purposes.

That there are sporadic attempts to modernise waterway cottages is undeniable, sometimes to a ludicrous extent whereby all that remains of the original after the rebuilding, at great cost, is one wall. Sometimes the local council will slap a demolition order on the premises and the tenant then moves to a council house; alternatively, if the structure of the cottage is good but the facilities are poor, it may be leased out, or even sold, to someone who is in no way connected with the waterway. Both these states are undesirable, for as each cottage is released there is less supervision of intermittent lengths of waterway.

The architecture of a local canal cottage in many ways typifies the attitude and finances of the company which originally owned the waterway. Nowhere is this more apparent than on the main line of the Grand Union Canal, where at the top of Bascote the old toll-house, now in private hands, shows only slight signs of its age in its small windows and soft red-brick shell; while at the bottom of the flight, within a mile, one of the 'new' houses, built by the Grand Union Canal Company in the early 1930s, an excellent house in many ways, nevertheless belongs to the 'suburban by-pass' school of architecture. On the Aire & Calder main line the lock-houses are usually built of stone, somewhat grimed now, and on the neighbouring New Junction Canal there are very attractive swing-bridge-keepers' houses, although as the swing-bridges have become unnecessary the houses are falling into private hands. The barrel roofs of some lock cottages on the southern Stratford Canal are quite well known, but the comments of one man who lived in one tend towards the unprintable, due to the inherent difficulties in decorating and that great bugbear of most canal-side premises – damp. The 'round-houses' of the Thames & Severn Canal and the neo-Corinthian houses of the bridge-keepers on the Gloucester & Berkeley Canal are also well known.

To say that a lock-keeper's job is to maintain his local flight of locks is both an understatement and is also inaccurate, for in all probability not only is he responsible for his locks but the towpaths above, between and below, the hedges, leak-stopping, bridge brickwork and all the other requirements of any well-run waterway. In time of flood, or during the winter, he might be up all night keeping an eye on the water levels, or he may be patrolling, in the face of the wind, some stretch of embankment that he knows to be weak, and it was in recognition of this that, until comparatively recently, a free issue of coal (upwards of four tons per annum) was given to the lock and reservoir keepers, in compensation for the fact that they would probably need a fire burning all through the night. Although the Truck Acts have outlawed the 'tommy shop' system of payment, there can be no doubt that in many ways the rents of waterway houses are kept at an artificially low level to compensate for the low wages paid. On the other hand if the house has been standing for 200 years its cost must have been paid many, many, times over.

At most entries to navigations will be found a narrow point or 'gut', often taking the form of a lock with, possibly, a building straddling the two sides. This, in one form or another – varying as it did from a primitive narrowing of the cut bank to a solidly-built block of offices – was the gauging, or bar, stop. Any traffic moving along the waterway could be charged a sum specified in the incorporating Act of Parliament, calculated on a ton/mile basis. Typical of these charges are those for the Barnsley Canal[4] shown below.

	d.
Wheat, Shelling, Beans, Pease Vetches and Lentiles, Rape, Line, Cole and Mustard Seed, Apples, Pears, Onions and Potatoes	6 Per Quarter for the whole length
Barley	5 ,,
Oats and Malt	4 ,,
Pack or Sheet of Wool, Dried Pelts or Spetches, Coal, Slack, Cinders, Culm, Charcoal and Lime	1 per ton, per mile
Limestone	$\frac{3}{4}$,, ,, ,, ,,

	d.
Stone, Iron-stone, Flag, Paving-stone and Slate	1 per ton, per mile
Pig or Old Iron	1½ „ „ „ „
Cast Metal Goods and Bar Iron	2 „ „ „ „
English Oak, Timber and Planks	1½ per Forty Cubical Feet per mile
Elm, Oak and other English Timber	1½ per Fifty Cubical Feet per mile
Fir, and other kinds of Foreign timber	1½ „
Deals and Battens, equal to Thirty Deals, of 12 feet long, 3 inches thick, and from 9 to 12 inches broad	1½ per mile
All other things not before enumerated	2 per Ton, per Mile

Each craft using a waterway was registered in a toll-book, which gave the draught of a boat when laden with *x* tons. Taking this loading figure as being an inch to the ton, allowing for the weight of the craft and its fittings, and adding a certain percentage for dunnage, the basic draught could be 12 in. A special rod or stick was always held which could be placed on the side of the gunwales, its bottom end being immersed in the water. It was graduated in inches. Thus our imaginary craft being shown to have a draught of 36 in. must, logically, have a load of 24 tons. At some points (the Worcester Bar in Birmingham is a very good example) manning these gauging locks took place around the clock, which necessitated houses for the clerks engaged in these duties. A variant on a gauging lock could incorporate a weighbridge so that the goods discharged from the boats into carts could easily be calculated. Sometimes these would be maintained by the canal company, particularly at a coal wharf where 500 tons of coal might be stock-piled at any one time. By the checking and weighing of all vehicles passing to and from this wharf pilferage could be kept to a minimum. It might even be that the canal company owned the stock-pile of coal and merely held it until required, hence on the Birmingham Canal navigations we find the notices quoted at the beginning of this chapter, and that below:

BIRMINGHAM CANAL NAVIGATION
Navigation-Office, Oct. 27. 1769

Wanted, for the Use of the Company of Proprietors of this Undertaking, a Number of Carts and Horses, for the Delivery, of Coals in this Town; and Horses and Men without Carts, the Company being already furnished with some of the latter. All Persons willing to enter the Contract with the said Company, will be paid either by the Day, or in Proportion to what Distances they draw, as shall be agreed upon. Apply at the Navigation-Office on Thursday, Friday, and Saturday, the 2d, 3d, and 4th Day of November next, between the Hours of Eleven and One, on each of the said Days.[19]

Another example of a canal-owned weighbridge, together with its tiny toll-house, is at Knowle Hall Wharf on the Grand Union Canal. Obviously, a craft newly entered upon a waterway had to be weighed as an entity, and a good example of a weighing dock may be seen at the Waterways Museum, Stoke Bruerne.

There were a number of fiddles carried out by boatmen and, occasionally, the company might require that the load carried in a craft be completely taken out and weighed on scales. In the case of the relatively soft, top-crop coal carried on some parts of the Birmingham Canal Navigations the complaints of the shipper and of the boatmen were vociferous, for the continual handling could mean the coal being converted into slack which brought a lower price on the market. In an endeavour to overcome this, various hydraulic machines were developed: 'Birmingham, July 23, 1770. We have the pleasure to inform the Public that the Birmingham Boat Company have invented a Method to weigh their boats by Tubes, and by an Experiment made on Tuesday last, they weighed by Scales 30 Ton 9 Hundred into a Boat, and weighed them afterwards on the Proprietors' Machine, and the Difference was only 11 pounds, which Invention is allowed to exceed any Thing of the Kind, and is an Incontestable Proof of the Justness of the weighing Machine.'[20] Another problem lay in the handling of bagged goods, and to this end a machine was developed and actually used, whereby a bag could be slung on the machine, a rachet moved one turn and a number would appear on the machine's indicator. If, assuming one were carrying sugar, a couple of

EARL
FELLOWS MORTON & CLAYTON LTD

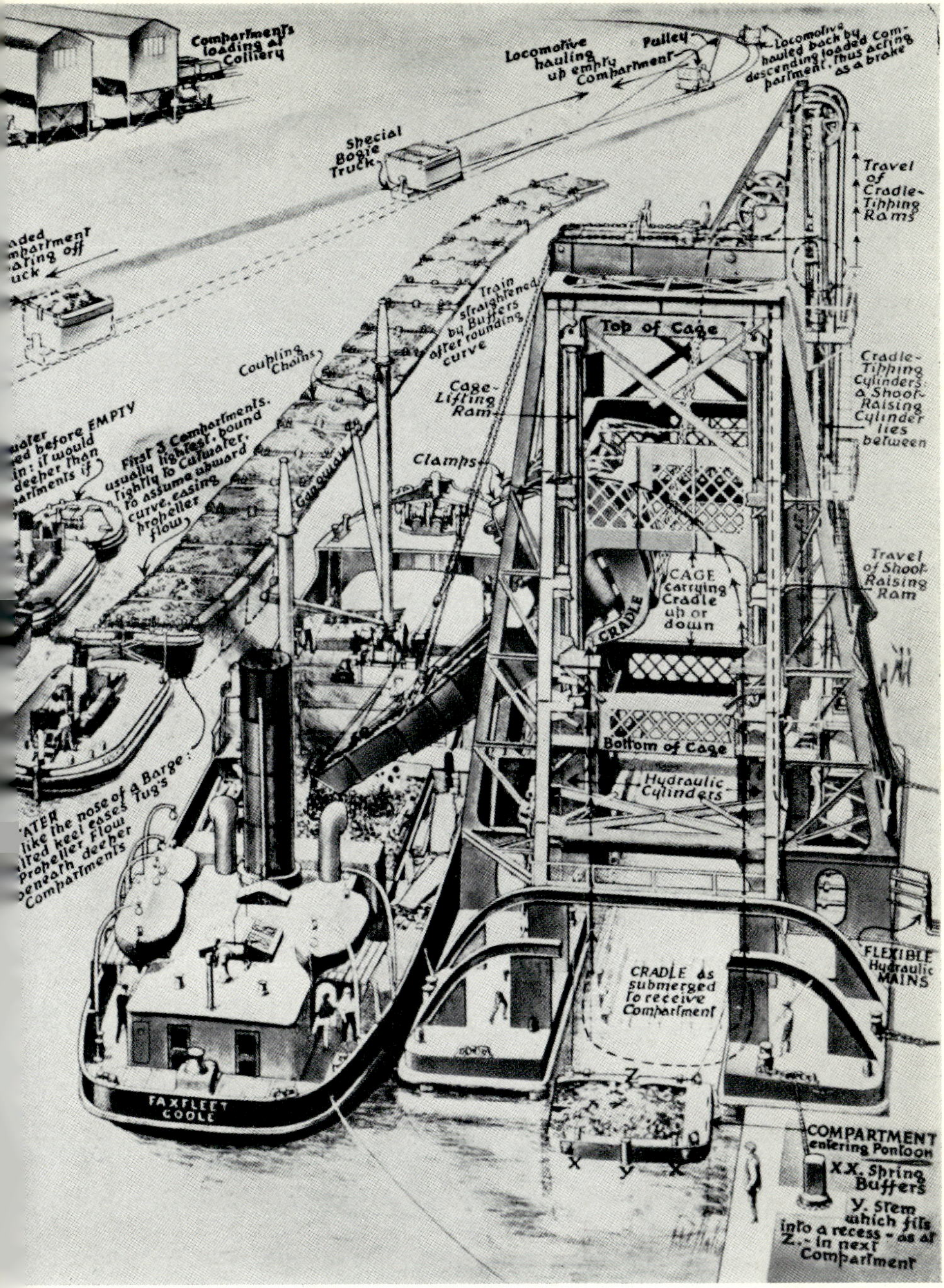

2 The gubbins for loading at Goole

23 Construction of a new by-pass bridge near Wharf Bridge, Aldershot, Basingstoke Canal, 1969

24 An early steam bucket dredger

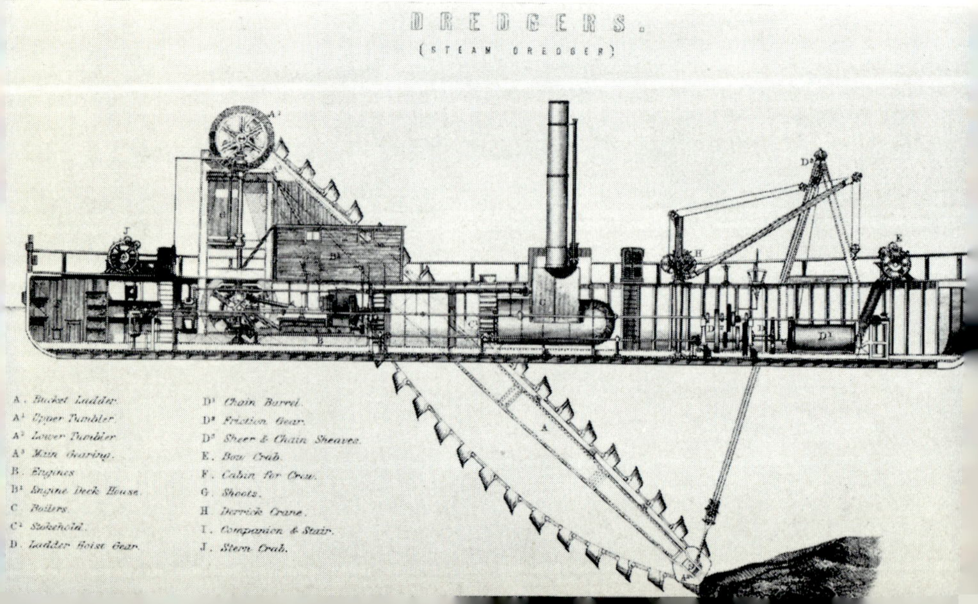

bags had gone astray *en route*, the whole art lay in persuading the weighing clerk to allow his attention to be diverted for a moment or two while a judicious application of the foot to the treadle or platform could increase the number shown to compensate. One had, of course, to ensure that one did not press the platform too often. Some coal-wharves tended to be situated in relatively quiet parts of the countryside and it was not in the least unusual for a load of coal to be discharged overside, the toll clerk taken away to the nearest hostelry by the boatman and, in the meantime, his mate would shovel back on board sufficient for their immediate requirements. It would be illogical to expect a boatman to buy coal, and until the virtual cessation of commercial traffic a certain amount of bartering inevitably took place between passing boats; one perhaps having flour or purée and the other having coal, although no one seemed amiably disposed to the boatman carrying fishmeal.

Wharves, often of great interest to the industrial archaeologist, represent in many ways the history of the waterway but most, with the coming of pleasure-craft and their demands for moorings, toilet facilities, car parks and cafés are now battered out of recognition. The humble farmer's wharf, usually a small wooden staging, has often been eroded by the effects of wind and rain; such wharves dated back to the days when many farmers were one-boat owners, sending away milk in churns, butter in crocks and corn in bags. Cadburys had a number of especially fast dairy craft whose sole occupation in life was to run along the Shropshire Union Canal picking up such churns of milk and butter, discharging them at Tyrley Wharf. A breakdown of carriers functioning on the Worcester & Birmingham Canal around the turn of the century shows nine boats in service owned by farmers whose cargoes were shown as 'farm produce'. On the Grand Union Canal about the same time there were a number of market boats which would pick up fresh vegetables from within a ten- to fifteen-mile radius of London and take them to the markets. Milk, euphemistically described as fresh, was brought by the same means and before we scoff at the time taken, say 12 hours to reach the consumer, it is salutary to remember that little, if any, of our domestic milk today is less than 48 hours old when we receive it.

The next type of wharf, of rather more durable nature, is or was that used for handling bricks, timber or similar heavy goods. Of necessity a portion of land would be bought, or leased, from the waterway company and a stone or brick curbed storage area built alongside the navigation. These items did not, in general, require covered warehousing but necessitated the building of a road. Sometimes this could lead to the wharf being built alongside a lock, which must have been a nuisance, and at other times it became necessary to cut away part of the bank to make a winding or turning place for the boats used.

The next extension to such a wharf would be a short stub arm, leading off from the canal. Such arms are commonly to be found, sometimes being used as the local rubbish tip. An interesting example, now disused, is the old Brewery Arm on the Stratford-on-Avon Canal near the Alcester Road Bridge. Originally leading to a malthouse it was subsequently used for the carriage of coal and, to a certain degree, of timber, but ceased to have any purpose in 1900; it is, nevertheless, more or less intact. From such a development would come the necessity for warehousing. As trade picked up this might lead to the growth of a town or village around the place where the wharves were first made. 'Preston Brook, is a hamlet at the foot of the hills, a little south-west from Preston, where the Trent & Mersey and Bridgewater Canals form a junction, it being the place of transhipment of agricultural productions and other goods to Liverpool, Manchester and other places. Extensive warehouses and offices for the different carrying companies have been erected, and a considerable extent of business is transacted here . . .'[21]

The next stage, as the demand extended still further, would be for a boat-building and repair establishment to be set up. In turn the canal company, who might only have kept a small office hitherto, would find it desirable to extend their scope of business, as on the Trent & Mersey Canal where the Etruria Wharves grew from nothing until by the 1830s:

> The various public wharfs, at which the raw materials are landed, and from which the manufactured goods are principally forwarded, are situate on the Grand Trunk Canal,

> near Etruria (which are called Etruria Wharfs) and near the lower extremity of Shelton (which are called Shelton Wharfs). They have spacious warehouses, sheds and accommodations, and are held by various carrying-companies. A railway, for horse-draught only, extends from Etruria Wharfs to near the middle of Hanley, about two-thirds of a mile. The principal Wharfs here are occupied by the following carriers or firms: Hugh Henshall & Co. (The Company's own carrying-firm), James Sutton & Co., Heath & Son, Morris, Herbert & Co. and (at Etruria Vale) The Anderton Carrying Company; Shelton Wharfs are occupied by Kenworthy & Co., Mills & Co., and Ebbern & Sons. A large timber-yard and boat-building establishment here is held by Timothy Dimmock & Co.[22]

The docking of boats for repair or servicing can be carried out in a number of different ways, the simplest being to utilise lines of railway track laid at an angle from the water's edge to a large open-fronted shop, the boats being drawn up with winches, broadside, and being rolled over manually if any work is required on the bottoms. An example is still to be seen above Nechells Shallow Lock on the Grand Union Canal. The second approach is to use an arm running off the canal, fitted at one end with stop planks, the boat resting on bostocks or stands above the bottom of the arm. After the stop planks or gates have been put in a closed position, a paddle is drawn allowing the water to run away down a culvert. This system is simple but wasteful on water. A later approach to the subject is to be seen at Tardebigge, where the water is pumped from the dock back to the level. Prior to the advent of electricity it is improbable that the cost of running and maintaining a steam engine for this purpose would have been economic. Yet another system calls for a boat to be drawn into a covered area on rollers or slides and placed in a special jig which enables the boat to be rolled over bodily to a 45° angle, the whole craft being suspended within a steel cage. This has the advantage that the whole of the bottom is easily accessible, no part being obstructed by the stands as in a normal dock.

Obviously, maintenance yards on waterways had a vital

function to perform. Until comparatively recently it was normal practice for lock-gates, stop planks and any iron-work required to be made locally and it is sad to reflect that the majority of the old maintenance yards are being leased out to outside interests, often not in any way connected with waterways. In some ways this is logical, as it brings in a greater income and waterways must be made to pay; but whether the extra costs involved in buying materials from outside contractors or of hauling them from the nearest main workshops are justified is arguable. Quite often too, these maintenance yards will show the growth and depth of commercial carriage on any one waterway, the oldest building may well date back to the early 1800s, with later additions. They are, almost invariably, dated by specially cast bricks or engraved stones and the dates continue through to as late as the 1920s, these later buildings being built, more often than not, of matching materials howbeit in slightly different styles. Any modern asbestos, concrete or aluminium structures are probably no connection with the canal company. Too often though, the fittings of maintenance yards remain medieval. In the 1950s, when cranes with wooden jibs went out of fashion, they would often be replaced not by modern electrical devices, but by even older iron ones, manually operated. Access to these yards, which dates back to the day of the horse, can prove very difficult with modern articulated vehicles and, indeed, at some it is necessary to specify the overall length of a vehicle to be sent.

In such yards the ghosts of commercial carrying are still about and most men tend to talk of the 'good old days' when there were 30 or 40 men employed, and busily employed, making and maintaining items for the waterways. The old men also tend to conjure up ghosts: 'I remember old Bert 'iggins in 1810 building a bridge on his own in seven days'. Such improbable stories are common even if real ghosts are not. Any logical person states point blank, and believes, that there are no such things as ghosts, but true canal people are illogical. Most firmly deny that there are such things, and then go on to tell you something odd that happened. The following was quoted, in a perfectly serious book, dated 1860.

'Well, gaffer, years ago, in the werry middle o' the tunnel, right between Tunstall on the one side, and Kitcrew [Kidsgrove] junction on the other, two men murdered a woman and threw her body inter the tunnel, and because it wor a deed o' violence, and her life taken from her afore it wor axed fur, that ere 'oman have never lain quiet; but years ago as it wor, she'll appear, sometimes in the form o' a white horse, sometimes like a female without a head; but whenever her comes, troubles sure to foller. Never wor there an accident at the collieries, but the Kitcrew bugget wor sure to come first to tell o' it. Somebody'll die, or be murdered, or drowned in the cut or coal-pit, when that 'ere ghost appears.'[23]

Another story concerns a boatwoman who, it seems, had been a terror to boaters all her life due to the awful language she normally used and who also 'continually indulged in drunkenness and just about everything that was bad'. One day in the year 1875, she began in her frenzy to swear and actually foam at the mouth. She took hold of the tiller and laid it on the deck and when she turned around to swear at the horse the tiller suddenly vanished out of sight, leaving no trace. The canal was searched but to no purpose, and from that day to this it has not been seen or heard of, the boatwoman believing that 'an angel or some spirit filched it'.

Any boatman worthy of the name will tell you the story of 'Spring-heeled Jack' and we relate it as it was told to us many, many, years ago.

'Going back a bit, ain't we my dears, but only to my father's time, when he was working down what was then the Grand Junction Canal. Now you know Cassio [Cassiobury Park] well enough I expect, and you know the Park Lock. Well near there was an old oak, and it happened during a storm this old oak got struck by lightning and me Dad did tell as there were nigh on two hundred keys [windlasses] dating back a long time in the middle of the tree, because yer see it was hollow. It seems the Lord who owned the land at the time when they first thought of building the cut, had in his employment a young and very athletic buck-nigger, and when the canal was dug this Lord did not want it, or the boatmen, and what he done was to have this nigger knock any boatman returning to his boat at night, on the head. They had usually had a pint or two, so the nigger easily took their keys and dropped them in this old oak, which caused the boatmen some difficulty when they wanted to work the lock the next day. Because they was a bit dazed from the beer and the clout over the head, and because they couldn't see him at night, him being as black as the very night itself, they thought it was a ghost as done it, and any that was brave enough to chase him never did catch him because he was so athletic.'

Superstitions regarding certain craft abound. When the boat No. 471 sank on the 'Moira Cut' (Ashby-de-la-Zouch Canal) in the early 1900s, the boatman who was inside the cabin was drowned and although the boat was subsequently raised no one would spend a night on it and it was broken up shortly afterwards. The butty boat No. 211, ex-Grand Union Canal Carrying Company, does not seem to have brought anyone any luck, there having been a couple of inexplicable drownings from her. Referring again to the 'Kitcrew bugget', we asked one old boatman who had worked the Trent & Mersey Canal all his life whether there was a ghost in Harecastle Tunnel, his answer was simple and to the point: 'No, no ghost but I gets down in the cabin and lets me mate [his wife] steer the boat through – just in case!.'

4 *Boats on a Canal*

Date when Gauged, Weighed, and Measured	21. 9. 00
	Tipton *Station*
BCN Register No. 17414	
Open Iron *Boat*	*Name* *No.* 74
Owner Patent Shaft & Axletree	*Address* Wednesbury
Extreme Length 70·7	*Extreme Width* 7·1½
Stowage ,, 62·4	*Stowage* ,, 7·0½
Draught when light 8·50	*Draught when laden with* 38 *tons* 44·10
Articles on board when weighed	5 beams, rudder, mast
	Cut up 10. 5. 27

Date when Gauged, Weighed, and Measured	25. 9. 00
	Smethwick *Station*
BCN Register No. 17423	*Name* The Empress *No.* 15
Cabin Wood *Boat*	*Address* Lifford
Owner G. Griffin	*Extreme Width* 7·1
Extreme length 70·7	*Stowage* ,, 6·7½
Stowage ,, 57·4	*Draught when laden with*
Draught when light 8·00	40 *tons* 45·92

Articles on board when weighed 5 beams, rudder, mast
Cut up 16. 3. 34

Date when Gauged, Weighed, and Measured 25. 9. 00
Smethwick *Station*

BCN Register No. 17425	
Cabin Wood *Boat*	*Name* The Linnet *No.* —
Owner L. B. Faulkner	*Address* Leighton Buzzard
Extreme length 70·11	*Extreme Width* 7·1
Stowage ,, 52·1	*Stowage* ,, 6·8½
Draught when light 10·10	*Draught when laden with* 38 *tons* 47·08

Articles on board when weighed 4 beams, rudder, mast & case, cratch, two stands, 4 planks, 4 stretchers, 2 barrows, 1 pump, 8 bags of corn, 2 cwt dunnage

Date when Gauged, Weighed, and Measured 28. 9. 00
Tipton *Station*

BCN Register No. 17433	
Cabin Wood Tank Deck *Boat*	*Name* The Derwent *No.* 14
Owner T. Clayton (O) Ltd.	*Address* Oldbury
Extreme length 69·10	*Extreme width* 6·11
Stowage ,, 52·8	*Stowage* ,, 6·7
Draught when light 12·40	*Draught when laden with* 36 *tons* 48·84

Articles on board when weighed rudder, mast & case, tank deck, 5 bags of corn, 1½ cwt. logs

Shown above is a random sample of craft types from a BCN Boat Register; 20 times this variety, all of which bar a few – tugs, maintenance boats and the like – were in use for carrying. Obviously some designs played a major part in the history of commercial carriage on waterways, others less strong numerically nevertheless had a duty to perform, perhaps the movement of a specialised item.

The most common heading in any Birmingham Register is 'Day boats without cabins', in itself a misnomer, for about

60 per cent of them had a 'hutch' which would hold four men and a stove (for use in inclement weather), but in contradistinction to 'cabin boats' were not (in theory) used for overnight sleeping or to accommodate families. A case was, however, reported in the *Birmingham Gazette* in 1794 when a man of 50, engaged in bringing coal from the collieries, returned to his boat against regulations and perished from exposure, the boat not being equipped with a stove or bed. He had had a number of drinks and death was occasioned, it was thought, 'by the rapid cooling of his body'. It was stated that he was in the habit of sleeping on the boat to save a night's lodging money.

It was these day or 'Joey' boats which then provided the greatest variety in craft design within one general classification, as they do today. One boat in the register, No. 16141, still remains in service with Messrs Stewarts & Lloyds at Coombeswood Basin on the Birmingham Canal Navigations. Of orthodox wooden construction, pointed at both ends, and measuring roughly 72 ft × 7 ft × 3 ft 6 in. over-sides, the minuscule cabin mentioned above has always been in danger of knocking itself to pieces in the low tunnel (Gorsty or Gosty Hill) at the entrance to the works, and provision is made, by means of a bung, for the boat to be partially flooded, thus increasing the unladen draught and reducing the freeboard (height above waterline). Obviously this increase in traction is advantageous to a tug driver – all this type of boat being engineless – as the steering is greatly enhanced, and the danger of becoming wind-bound (blown against a bank by the wind) eliminated. Iron boats of a similar class are even more sophisticated, being fitted with flooding taps and permanent plunger pumps. A further development of this flooding arrangement is to be found on some mud-hoppers used for the removal of waste from dredging sites. These have an internal tank with either a 'drop' bottom, for circumstances which permit the discharge of spoil at sea, or a raised bottom, making a double skin. Due to their inherent top-heaviness some form of ballast is desirable; water is economic and easily discharged by electrical pumps.

Nearly all wooden boats, whatever their construction, were fitted with bungs for two purposes other than those above. Firstly, should there be no traffic they would be taken to a quiet backwater where the bung would be knocked out and the boat allowed to sink. Like their iron and steel brethren the greatest wear is inevitably 'twixt wind and water', a controlled sinking on to mud is preferable to unwanted perching on stones which might break her back. Secondly, use of the bung could be made to drain water from the craft if she accidentally sank or was going on dock. Not too long ago an elderly day boat, laden with 35 tons of slack sank in the Farmers Bridge flight of locks on the Birmingham Canal Navigations, due to the dislodgement of a piece of oakum. Two alternatives existed after the slack was shovelled out. The boatman desired to have the craft lifted at one end and after removal of the bung, as the water drained out, to let her natural buoyancy assert itself – they will float half full – the leak then easily being stopped; but he was overruled by the canal engineer who essayed a straight lift which broke the boat's back. After this little could be done except to drag the boat out and break it up.

A second type of Joey boat, No. 16211, was listed as a derrick or crane boat, being fitted with a small manual winch amidships with a capacity, probably, of about five tons (three in later years). This boat, albeit with reduced carrying capacity, was in great demand, being still in service in a maintenance role, carrying piles, until 1951. Obviously it had many advantages, not the least being the ability to load and unload awkward cargoes at remote factories or wharves which were without mechanical canal-side facilities. Fishing unwanted objects from the bed of the waterway would undoubtedly fill in any spare time!

One functional advantage of a Joey boat was that, being double-ended, the helm could be put on either end, thus eliminating the time taken to 'wind' or turn a normal cabin boat.

Day boats with cabins could include a number of classes of craft, ranging from the short boats of the Leeds & Liverpool Canal to the Birmingham area station, or railway, boats. There is a curious difference in the life-span of a narrow and a wide boat, for example craft engaged on the Calder & Hebble Navigation in 1970 were none of them over 30 years old; some that were 35 years old suffered the ignominious fate of being dragged out of the water into a field and being burnt, others

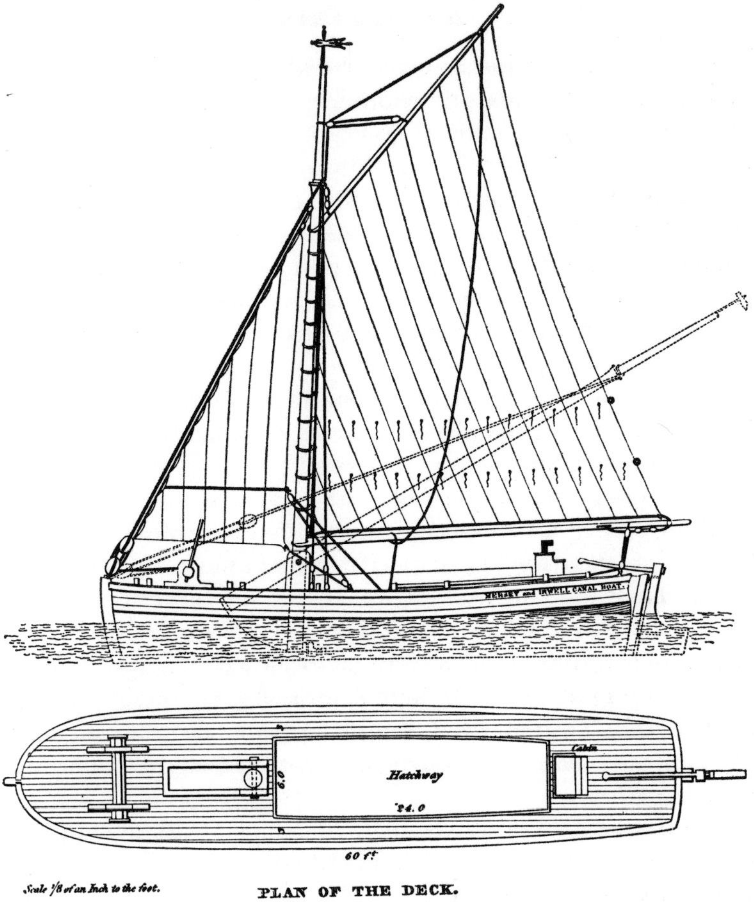

5A.

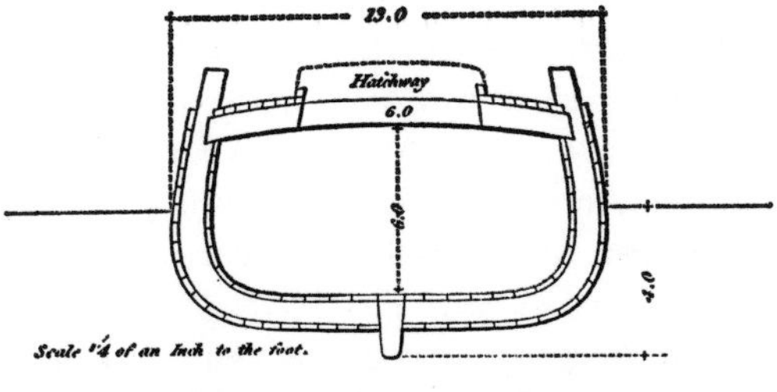

Figs 5A and 5B. Sailing flat, Mersey & Irwell Navigation, 1838.

have ended up serving as boys' clubs and floating restaurants. Wooden narrow boats, on the other hand, will have a minimum life, with reasonable care, of 50 years – spending those 50 years in hard commercial graft. Virtually all the cabins on these craft are equipped with stoves, or in more recent days with gas fires, and benches. Some, for reasons of economy, have steel cabins, which being unlined are not very popular due to condensation troubles and may be one of the causes of the rheumatic and bronchial complaints which often beset day boatmen in their later days.

On the majority of northern canals the cabins are capable of being dismantled, mainly to clear low bridges with the craft when running empty.

The railway or station boats have a long and honourable carrying career, although those that remain are often outclassed in so far as their tonnage capacity, a maximum of 22, is inadequate. Their fundamental use was within the Birmingham area, taking coal to the various railway yards wherewith to provision steam locomotives. Others were factory-owned, again carrying coal, running at relatively high speeds. As one might expect a horse-drawn station boat could operate faster and more economically than a diesel-driven motor boat on these short day runs.

Within the field of day boats with cabins could be found the market passage boats, which operated on a fixed timetable through various waterways. The best known, if only for their speed, were those utilised on the Glasgow, Paisley and Ardrossan Canal. In 1820 an advertisement for this waterway route appeared:

> A boat for the conveyance of passengers and parcels leaves Johnstone for Port Eglinton, Glasgow, every morning, Sunday excepted, at 9 o'clock, passing through Paisley at 10, and at 4 o'clock in the afternoon from Johnstone, passing through Paisley at 5. NB an additional boat starts from Paisley at 9 in the morning, every Wednesday [Glasgow market day] and returns the same evening at 6 o'clock in the summer, and 5 in the winter. Fares to Glasgow – 1st cabin 1*s*. 2nd cabin 9*d*. Trading Boats on the canal start from Paisley to Port Eglinton

at half-past-10 in the morning, and 12 at noon, to Johnstone at 11 morning, and from Johnstone for Paisley and Port Eglinton at 9 morning.

These packet-boat services were rather more extensive than is normally presumed. The Duke of Bridgewater was the pioneer, his services running from Warrington to Manchester, initially with two passage boats, one being a single-cabined craft and carrying passengers at one shilling (5p) each, the other being a saloon craft with three separate cabins, varying in price from 10*d.* (4p) to 2*s.* 6*d.* (12½p) per head '. . . and it is the pleasantest and cheapest mode of travelling you can conceive'; such indeed was the demand for passage craft of various kinds that even poets could wax lyrical, as did Mr Maxwell early in the 1800s, referring to the Forth & Clyde service:

> For here a cabin at each end is found,
> That doth with all conveniences abound,
> One in the head, for ladies nine or ten,
> Another in the stern, for gentlemen.
> With fires and tables, seats to sit at ease,
> They may regale themselves with what they please,
> For all utensils here are at command,
> To eat and drink whate'er they have at hand.

On this particular service in the course of 1812, 44,000 passengers took the boat from Glasgow to Falkirk, covering 25 miles in three and a half hours; by 1836 the figure for passengers was well over 200,000.

An amusing story is told of a mishap on the Worcester & Birmingham Canal in 1796. It is said that a party of Birmingham people, with a band of musicians, were travelling to a fair on a passage boat and were within sight of their destination when a rush of water overturned the craft. All their paraphernalia, including musical instruments and wigs, floated down the canal; however, nothing daunted, after the retrieval of both band and instruments, they marched to the fair and gave their recital. It is to be hoped that they were able to dry their outsides and wet their insides sufficiently well to compensate. While this accident was a frivolous one, that which took place on the Glasgow, Paisley and Ardrossan Canal in 1810, caused by some drunken oafs, was not, for at least 90 passengers were drowned. The *Flying Post* of 22 November reported: 'One family consisting of husband, wife and two children, were in the water. The husband is recovered, and a child, but the mother and the other child are gone. Another man has lost his wife and several children, and one woman has lost her husband and one or two children. We have indeed a general mourning. In one small street lie nine dead bodies. Some families have lost two children, some an only son!'

In many ways it is perhaps just as well that, other than for purely pleasure purposes, this type of working has come to an end. Sir George Head, a somewhat humanitarian man for the time, did not altogether approve of the horse-management on the Lancaster Canal and in his *Manufacturing Districts of England* published in 1835 he writes:

> The proprietors of these canal-boats have endeavoured to establish a theory, which, setting philosophy aside, is surely a bad one for horses; they maintain, that the animal works more at his ease at the rate of ten miles an hour, than at eight, or even less; because the swell at the head of the boat is, they say, by the greater velocity, surmounted before it accumulates, whereas at less speed the increase of the obstacle more than counterbalances the diminution in labour. Much depends, at all events, upon the width of the canal, the depth of water, and so forth; but, in practice, I think the experiment fails; I never saw horses more defeated than these, although the stages were usually only four miles. At the end of each they sweated and panted, as if they had undergone a severe burst with foxhounds; there they stood planted as it were, reeling and shaking their tails till led away. We were generally on these occasions very soon out of sight, for on changing the cattle no other ceremony was requisite than merely to unship the eye of each trace from the hook, and fix the other instead; nay, so quick were our movements, that frequently, on whisking round a corner, a traveller was seen waiting for a passage, and within the space of twenty seconds, from the

Hazelhurst Aqueduct, Endon, Caldon Canal.

REGISTERED AT
LEES & AT
POLESW

moment the boat stopped till she proceeded on her way – from the blast of the horn to the sound of the whistle, the packages and our new companion, the owner, were all together gliding away on our voyage.

Of cabin, i.e. residential, craft it is probably an understatement to say that there were manifold varieties, as the permutations were infinitely variable. The best known, if for no other reason than their exposure through the popular media, are the pretty, painted narrow boats of the midlands. Outclassed for general carrying their movements on the few remaining traffics that were suitable have been stultified since nationalisation. In all probability cabin boats as we know them today did not exist in the early years of waterways when there was little financial necessity for a boatman to take his family on board the craft; for he was, like the coachman and the modern lorry driver, one of an *élite* group, and a good boatman could be sure of finding employment almost anywhere.

The orthodox crew for a horse-drawn single boat would have been one lad on the towpath to guide the horse and an older man to steer the boat. Lock working was almost mechanised, the horse being trained to pull and stop at the required moments. In various conditions and positions he would act as accelerator, brake and clutch combined in one in addition to his main function as the motor.

The man and the youth would, on arrival at a staging post, place the horse in the stables and attend to his requirements (if there was no ostler). They themselves would stay at a suitable hostelry, although on some waterways sheds were made available by the canal company for their overnight accommodation. Their wives and children could stay at home in their cottage. Eventually as trade decreased so it became logical for the boat's captain to use his eldest son in lieu of the lad previously employed and when the son grew up and left to take his own boat there was room for the wife and younger children. For financial reasons the wife now found herself steering the boat as it was apparent that economics no longer permitted the running of what was effectively two households, and living on board became permanent. Unfortunately, it is axiomatic that the

25 A dislodged piece of oakum caused this boat to sink near Blisworth Tunnel, Grand Junction Canal, *c.* 1910

26 A pair of tar boats at Banbury, Oxford Canal, 1950

27 Compartment boats with steam tug. Aire & Calder Navigation, *c.* 192

28 *Freight Pioneer*, the first British Waterways Board 'push–tow' craft

larger the cabin the less the carrying capacity of the boat, and there can be little doubt that overcrowding did in some cases take place.

Mr John Harris of Falmouth, the 'Cornish Poet', in 1860 asked:

> Seven in a small boat huddled,
> With beds, and pots, and cans,
> When wintry winds are wailing,
> Or summer zephyr fans,
> Even from the year's beginning,
> Until it endeth slow;
> Ye who give laws to England,
> How long shall this be so?

But the statistics relating to the boatmen in the employ of the North Staffordshire Railway Company in July 1875 with two exceptions do not seem to bear out this rather general contention.

Name of boatman	*M or S*	*No. of children*	*If brought up in boat or not*	*Age*	
J. Butteress	M	6	Brought up in boat	38	Cannot read or write
S. Wilkinson	M	1	,,	50	,,
J. Curtiss	M	1	,,	45	,, (child)
W. Taylor	M	3	,,	30	,,
W. Griffin	M	2	Not brought up in boat	26	Can read and write a little
O. Wooding	M	1	Brought up in boat	50	Cannot read or write (child)
S. Hodgkinson	M	1	,,	33	Cannot read or write
W. Clews, Sen.	M	7	,,	43	,, (excepting one child)
A. Jinks	M	–	,,	50	Cannot read or write
W. Clews, Jun.	S	–	,,	21	,,
J. Curtiss, Jun.	S	–	,,	23	,,

However, due to pressure by various 'do-gooders', an Act became operative on 14 August 1877 for the registration and regulation of canal boats used as dwellings.

> After the expiration of twelve months after the commencement of this Act, or if the regulations of the Local Government Board hereinafter mentioned have not at that time come into force, then after the expiration of six months from the date at which they have come into force, a canal boat shall not be used as a dwelling unless it has been registered in accordance with this Act. The owner of a canal boat may register that boat with the registration authority hereinafter mentioned as a dwelling for such number of persons of the specified age and sex as may be allowed under the provision of this Act; and the boat shall be used as a dwelling only for the number of persons of the age and sex for which it is registered.

Descriptions of narrow-boat cabins abound; usually omitted are the two items which will be found and accepted as essential on any pleasure craft today: toilet facilities (a bucket in the engine-hole being considered good enough for boatmen) and water capacity – for the boatman, two cans on the roof.

For various reasons cabin boats were never widely used in the north, probably the main reason being that tonnage capacity has always been greater and the wages higher. A modern example of the effect of low wages upon craft relates to the boats *Maudie* and *Ada*, both iron cabin boats and now lying on the bottom of the Basingstoke Canal. When offered for sale in 1906 they were bought by a man and collected from Nateley where he took his wife and his children. Living in the cabins they began to bowhaul the boats down the canal but, unfortunately, a stoppage took place and they had to wait. Whilst doing so they contracted diphtheria; the children were taken to an isolation hospital in Guildford and the parents disappeared, leaving tolls and dues which they could not now pay behind them – their capital had been expended and the enforced layover left them with no money. Eventually the craft sank and have remained so.

It is not our intention here to give technical descriptions of boat types – these may be found elsewhere[31] – but it is not inopportune to quote a description of how typical craft were built on the Basingstoke Canal at Ash Vale during the period 1918 to 1949. Barges were built on the boathouse side of the canal, repairs being done on the other side, because it was easier to chain winches down to the trees for winching out operations. Barges came off the production line at approximately one-year intervals and there were two main types: 'Reso's' and 'Oddn's'. abbreviations used to indicate residential boats, accommodating two people, and odd barges with no accommodation, used on day work. Most had transom sterns and would load 75 to 80 tons on the Thames, carrying 50 tons to Woking; they were built of English oak with Columbian pine bottoms. The construction, to Mr A. J. Harmsworth's personal design, embodied many features which had proved to be sound practice, giving strength and long service, and they were all moulded chine (i.e. two turn-up planks) which made the barge easier to steer in the confined channel of the canal. With bottoms of 2½-in. thick pine and 1-in. deal sheathing inside, the chines were 3-in. oak, the sides 2-in. oak with ½-in. oak sheathing inside, frame 4-in. by 4-in. oak, crooks or knees 4-in. by 7-in. and floor timber 5 in. by 5 in. Overall they measured 72 ft. 6 in. with a beam of 13 ft. 10½ in.

According to Mr Harmsworth the distinguishing marks were the almost straight stems and booby hutch cabins with 12-in. thick wash-board for'ard and 9-in. wash-boards aft. Most had iron bollards, one on each quarter. The building started with the laying down of the bottom planks inside the corrugated iron barge-building shed, deal sheathing was then nailed across and floor timbers spiked down, stem and stern posts erected, followed by the knees, timberheads and crooks. Next two turn-up planks were fitted and the sides planked up, steaming to give the requisite curve where necessary. Once this was completed the gunwales, coamings and decks were put on, the flooring in the hold put in (the false floor) and the fitting-out done. Some tar and paint and the barge was ready, except that water was pumped in on several consecutive days before the launching in order to swell up the bottom planks. The barge was jacked up, skids were placed underneath and greased. On the day of the launch the barge was lowered until it just rested on the top of the skids and before the jacks could be removed two struts – one at the head, one at the stern – were put under and the whole weight of the barge taken on these. On the count of three the struts were simultaneously knocked out by men with large iron bars, the barge dropping down on to the skids and sliding into the water broadside on. After launching the new barges were taken across the canal to have rudders fitted, which was done by dropping them on with a chain hanging from a tree branch. The final touches, such as the fitting of hatches and supplying of tarpaulins, were made at this point. Mr Harmsworth says that the whole operation would have been completed within four months.

The basic causes of the decline in residential craft are not difficult to establish. Disregarding any official attitudes that exist, the conditions in cabins are no longer acceptable to women. Even if they are prepared to forgo their television, washing machine and other domestic appliances, they will not willingly live on almost non-existent wages. The average take-home pay for the captains in 1963 was something to the order of £8 per week with wives and children giving their labour free. During the British Transport Waterways régime, tonnage rates were calculated for all trips on the narrow canals, and a few are quoted below; a maximum payload for a pair was 55 tons and overall working time averaged by then not more than two-and-a-half linear miles per hour.

From	*To*	*General goods per ton rate* *s. d.*	*Total rate inc. 95% cost of living bonus at 7th March 1960* *s. d.*
Brentford	Tring	1. 11.	3. 3.
Brentford	Wellingborough	3. 1.	6. 0.
Regents Canal Dock	Leicester	3. 8.	7. 2.
Regents Canal Dock	Brentford		2. 0. nett

From	*To*	*General goods per ton rate* s. d.	*Total rate inc. 95% cost of living bonus at 7th March 1960* s. d.
Bournville	Brentford	3. 10.	7. 6.
		Bowhauling 2. 6.	Bowhauling 5. 0.
Birmingham	Nottingham	2. 7.	5. 0.
		Bowhauling 5. 0.	Bowhauling 10. 0.
Southam	Birmingham	1. 5.	2. 9.
			Fly money 6*s.*
	Coal per ton		
Coventry and Newdigate	Dickinson's Mills	2. 5.	4. 9.
Griff	,,	2. 6.	4. 11.
Baddesley, Orchard and Mancetter	,,	2. 7.	5. 0.
Amington, Pooley Hall and North Warwick	,,	2. 9.	5. 4.
Measham	,,	3. 0.	5. 10.
Northampton	Greenford (*coke*)	2. 4.	4. 7.
	Light running payments per boat		
Birmingham	Amington and Pooley Hall via Fazeley	8. 4.	16. 3.
		Bowhauling 5. 0.	Bowhauling 10. 0.
,,	Amington and Pooley Hall via Braunston	15. 0.	29. 3.
,,	Measham via Fazeley	18. 0.	35. 1.
		Bowhauling 5. 0.	Bowhauling 10. 0.
,,	Measham via Braunston	15. 0.	29. 3.
Hawkesbury	Weston Point	26. 8.	52. 0.
Croxley Mills	Measham	45. 6.	88. 9.
Leicester	Bulls Bridge	33. 4.	65. 0.
Fly Money (General)			20. 0.

Fly working here means working the clock through, although as most flights were padlocked at 7 p.m. this was obeyed more in spirit than in the letter. One boat captain, lying at Broad Street Wharves, Wolverhampton, was told to go forthwith to Anderton Basin on the Trent & Mersey Canal, a mere matter of 82 miles and 54 locks, which journey he performed, admittedly with only a single motor boat, in 23 hours. On arrival he staggered up to the superintendent's office only to be told 'Oh, the goods have gone by lorry!' Tut-tutting philosophically, the captain went back to his boat and slept.

There is a tendency in modern economics to find that a tug plus a number of towed or pushed dumbcraft, with the boatman going home by motorbike every night, is far better than residential boating. About a hundred years ago the Aire & Calder Navigation, faced with the severe competition of railways, instead of buckling under made an all-out effort to compete by dredging, lock-lengthening and building up the facilities at the port of Goole. All assisted, but the key factor was their engineer, William H. Bartholamew, for not only was he a good engineer but also a man of fortitude and vision. Small wooden tub boats had been tried on various canals, but they had only been mediocre in practice, probably because the horse lacked suitable power, and therefore no great future had been envisaged for this method of transportation. The Aire & Calder Navigation had thousands of tons of coal to transport, and a method of moving vast quantities in bulk, quickly and cheaply, would be of great advantage. Steam tugs were now making an appearance, and these, utilised properly with an efficient tub boat, could be the answer to all their problems. The earliest tub boats (more popularly known as Tom Puddings) of the Aire & Calder Navigation made their appearance in 1860 and were able to carry between 10 and 15 tons each. The locks on the canal had been enlarged to a length of 215 ft and were 22 ft wide with a depth of 9 ft. With the size of the locks so greatly improved it followed that the Tom Puddings could be of a large size and were developed until they became a steel construction, 20 ft long and 15 ft wide and able to carry between 35 and 40 tons of coal each.

The ends of each compartment boat were designed so that

each boat was cunningly attached to its neighbour. While in motion no side movement occurred howbeit it was still possible to navigate around bends in the waterway. Towed by steam tugs in trains of 19 the total average tonnage moved per trip was 740. William H. Bartholamew received the satisfactory sum of $\frac{1}{2}$*d.* per ton royalty for all coal transported by this method. Continual progress was made, regular trains of Tom Puddings made their journey to the port of Goole. This unique compartment-boat system and shipment of coal, along with the rise of the Goole Steam Shipping Company, brought the port into prominence as one of the more important ports in the country, despite the fact that it is 50 miles inland from Spurn Point. Five 52-ton Compartment Boat Hoists were erected at the port, one of them being mobile. These hoists of gigantic proportions lifted a loaded compartment boat out of the water and tipped its contents directly into the holds of a ship. Because of age the first hoist was dismantled in 1920, the second to be dismantled was removed as recently as 1969. The remaining three hoists, one of which is mobile, still survive, and are still used daily. In 1969 one of the hoists was given a general overhaul to ensure its satisfactory working life for many years to come.

From the early days a towing piece was used, called a Jebus, to eliminate some hydrostatic problems. It is a specially built bows-only section of a barge and is pushed in front of the tug when the tug is towing empty compartment boats. When the boats are loaded the Jebus is towed in front of the train, behind the tug. This enables the water pushed back from the screw of the tug to separate before the loaded compartment boats. Without this device the action of the water pushed back by the tug against the wall of steel of the first compartment boat, which when loaded extends some 7 to 8 ft below water level, has some adverse effects on the handling and propulsion.

Tom Puddings were, and still are, loaded by gravity chutes, with one exception of note. Coal from St John's colliery, Normanton, some two or so miles from the navigation, was loaded directly from the colliery screens into the boats. They were then transported on special railway trucks to and from Stanley Ferry Basin. The trucks with the compartment boats on top were lowered down a slipway into the water, where the compartment boat floated off, or on if empty, to be pulled out for loading. A steam tank locomotive was used both for transporting the trucks and compartments to and from the colliery, and, by means of a system using pulleys and ropes, for trucks using the slipway. During 1942, subsidence had adverse effects on the railway line in the vicinity of the colliery, and as the expense of making good was considered prohibitive the practice ceased.

British Waterways, who took over the Aire & Calder Navigation upon nationalisation, continued the use of Tom Puddings, new diesel tugs named after Yorkshire collieries being obtained to replace the former steamers. Seven tugs are now in use with the compartment-boat fleet, six for towing trains and one permanently at Goole used for 'shunting' operations at the hoists. The number of compartment boats in use over the years has fluctuated, the total now in service being reduced to 450. Compartment boats in the past have, besides being used on the Aire & Calder Navigation Main Line, been used on the Barnsley Canal (abandoned 1953) and the Calder & Hebble Navigation. In the case of the latter the last compartment boats were used in January 1966, for it was decided to concentrate activities at Castleford, and therefore coal now goes by road to Castleford and is transhipped there into the compartment boats.

Loading now takes place on the Aire & Calder Navigation at Castleford, Woodlesford, Parkhill, Allerton Bywater and Kellingley and on the Sheffield and South Yorkshire Navigation at Doncaster with coal ex Maltby, Yorkshire Main, Bullcroft and Cadeby Bossington, and coalite in small quantities by road from Askern Bolsover. The tonnage of coal transported to Goole for shipment in 1968 totalled 530,000. Most of the coal is shipped coastwise for gas and electricity undertakings, and foreign to Scandinavia, France, Belgium and Holland, with small cargoes finding their way to Malta, Gibraltar and the Faroe Islands.

In quite recent years the compartment boats were built and repaired by the British Waterways Board at their depot at Goole, but since April 1968 the Board have only undertaken minor repairs; all building and major repairs have gone to contractors.

By contrast the Leeds Industrial Co-operative Society still

hold the distinction which no other retail Co-operative Society can possibly equal. They are the proud owners of a fleet of canal barges and a waterside depot. Unfortunately their traffic, with the coming of smokeless zones, has declined, but since the 1930s they have gradually eradicated their standard barges replacing them instead with tugs and dumb barges.

Nowhere can the death of barges and boats be better shown than by the statistics of the number of barges passing under Royston Lift Bridge on the Aire & Calder Navigation en route to the Barnsley canal:

1943	*1944*	*1945*	*1946*	*1947*	*1948*	*1949*	*1950*
856	674	542	703	124	54	15	4

Nevertheless, the outlook for waterways in so far as commercial carrying of this type is concerned is not entirely gloomy. In 1967 the British Waterways Board considered replacements for their rather elderly craft in use on the Sheffield & South Yorkshire Navigation. It was not felt that individual 100-ton barges, each requiring one or two men as crew, would be economic and instead, at the end of 1969, orders were placed for nine compartment boats, each of 150 tons capacity, which can be pushed two at a time by a new type of tug. To give manoeuvrability an outdrive system, the German Schottel propulsion unit, was utilised. The first was launched in September 1970 and, after trials, she delivered her first consignment in December of that year.

This new tug is, functionally enough, called the *Freight Pioneer*; a far cry from those names of early craft – *Rosa Ann*, *Elizabeth*, *Rachel* were probably names of wives or children, others, *Independent*, *Perseverance*, *Unity*, typified their owners' ambitions or philosophies with *Live and Let Live*, *Chance*, or *Look Up*! Flowers too, were popular, *Tulip*, *Rose in June* and *Marigold* were pretty enough for a 'rough old barge'. Battles – at least those we won – and the royal family had their share, *Waterloo* and *Princess Mary* are but two. Towns and rivers were extensively used, sufficiently so for two classes of craft to be named after them, but they were not always very imaginative, *Bawtry* and *Bognor* vie with the *Diglis* and *Smethwick* for prosaism. One likes the idea of a boat called the *Hope II*, *Lord Babs* or the odd *Algebra*. The succinct *Bee* conjures up visions of a light fly boat while the *Happy Return* must have had a cheerful enough crew. Did the captain of *Nelsons Valour* take heart when iced-in one January in 1909? Seemingly so, for this wooden fly boat, built in 1894 was still about in 1971. Like seamen, canal people prefer not to change the name of a boat, although a modification as with *Ivy* to *Hedera Helix* is permissible and indeed *Why Not?* Certainly *Willie and Bertie* will not mind, although the *Imp of Tradition* might object!

One little-known episode in the lives of some craft resulted from their being dispatched from the green fields of England to less salubrious parts of France during the First World War. One eyewitness, then Sergeant (Inland Water Transport Corps, Royal Engineers) later Captain (M-N) A. V. West stated in a letter to Turnpike Sailors Ltd that:

> 'In 1915 with IWT tug As88 I towed two narrow boats from Mills & Knights Dry Dock, Rotherhithe to Calais – from there to Abbeville on the Somme – we had a very rough passage and I was pleasantly surprised what good little sea boats they were. Later on in 1916, I was detailed for a special job of towing floats of high explosives up to the batteries on the Nieuport-Dixmuide front with the *M.L. Geranium* – it was then I met those narrow boats again, they used to go up a small canal that ran through a forest, and load great heaps of thickets for trench supports. The IWT nicknamed them the *Robin Hoods*. I can assure you they did a fine job. I received a special commendation – and so did your NCOs in these advanced front line jobs. The engine was on the top of the cabin. Although it was very hot up the line at times, I think we all preferred it away from the discipline of the base. As long as I got my cargoes to the guns I was left alone!'

One wonders where the boats ended up. Other canal craft were later to be used, during the Second World War, on the retreat from Dunkirk. At least one is still extant, proudly bearing a bronze plaque to commemorate the occasion.

5 *Crews on Boats*

> In this year 1783 . . . The flame broke out at Etruria. A boat laden with provisions stayed at the wharf there, and the goods, as it was supposed, were for use in the Potteries. But as soon as it was found they were destined for the Manchester market, this simple fact was interpreted into a design of the owners to enhance the price and scarcity of provisions. Word of this was conveyed to Shelton and Henley; and the result was the collection of a vast mob which hastened to the scene of action. The boat meanwhile had gone on its way, but it was followed to Longport, seized, brought back to Etruria and with another boat which had just come up the locks, was rifled of its contents. These were sold at such prices as the mob pleased. An endeavour was made to give an air of justice to this daring act, by handing over the proceeds to the captains of the boats; but the real spirit of the rioters was shown in the attempt made to rifle and fire several houses to other mischief . . . Two men were subsequently arrested . . . Both were tried and convicted, and one named Stephen Barlow was executed.[24]

A rather grim experience for the boat's captain, but to his credit he, who could so easily have joined in, tried to resist.

During the general strike of 1926 a number of day boatmen found themselves unwilling accomplices of the strikers. Small groups of colliers in Birmingham had leased plots of land at £1 a time and making a 'winning-hole' or 'bell pit' extracted as much coal as they could, selling at 18*s*. (90p) per ton to the boatmen. Obviously this black-legging was regarded with an unfavourable eye by the strikers. Initially the boatmen shovelled ashes and dirt from the towpath on to the load, giving the impression of carrying rubbish, the said tactic also permitting a lower rate of tolls. Once wise to this subterfuge the strikers lay in wait one day at the top lock of the New 'uns (Farmers Bridge 13) and, turning the horses loose, gave the boatmen two alternatives, sell the coal or be sat on. Some sold, some swam, the upshot was the same – the coal was sold for what it would fetch and the proceeds were given to the boatmen.

One day in the middle twenties a boat-owner found that the flight of locks he wished to traverse was closed for an emergency stoppage. He thereupon decided to lay up in the private basin above the top lock. 'Sorry, Mr Smith,' said the owner, 'you can't stay here.' Not unnaturally, for he knew the man well, Smith asked him why. Jerking his head the owner walked around the corner of the warehouse and asked Smith how well he knew his butty (mate). 'He's good,' said Smith, 'he don't go a-baccering [idling] along the towpath, but keeps a good eye on the horse.' The owner grunted, 'Leave your boats if you must, but tek him away – he's the biggest rogue on the cut, used to work for me until he sold me horse, boat and harness. Got three quid for the lot, and nine months after, but I dain't get anything back.'

In 1910 the father of a friend of ours wanted to wed a lass. He was a day boatman and born in Pelsall and she was a miner's daughter from Walsall, only six miles away. Their wedding took place with an armed (bludgeons and staves) guard of boatmen to protect them. Five men were arrested including the father-in-law who openly stated in court that he wasn't going to have his well-brought-up daughter marrying a boatman. About the same time a long-distance boatman, operating between London Docks and the Northampton Mills – wheat in and flour out – fell in love with one of the miller's daughters. There were 60 guests at the wedding and afterwards a new boat showed the attitude of this father-in-law. So whence comes the disparity?

It is usual in this day and age to regard boatmen as rather 'quaint' and a lot is talked and written about their honesty, uprightness, illiteracy, their hard life, their clothing and so on *ad nauseam*; something to be stared at – 'Fancy living in one of those cabins, how did they do it?' – something to be treated as a pariah – 'Come away from those barges, they're dirty!'

The simple truth is that bargees and boatmen were from 1760–1860 basically much as any other people of the labouring classes, with a few of their failings and more of their virtues. They had as much to put up with after 1860 as a 'tied' farm cottage tenant, for, too often, they did not own their own boats but worked for a 'boss' without union or other protection. On the day boats life was harder for the captains than for the long-distance boatmen in terms of conditions and the work involved to earn a day's pay, but far easier for the family as they had a roof over their heads. In the same context it must be admitted that unless the boys of the family joined their father boating they had to go to work in a mine or factory as early – in 1846 – as nine years old. By contrast the family of a long-distance man automatically started work as soon as they were old enough to push a gate or lift a paddle and most had only one ambition – if a girl, to marry a boatboy, or even a canal man 'on the bank' (usually a lock-keeper); if a boy, to have a better boat and horse than his father.

There was a certain degree of interchange between men on the bank and men on the boats and, in general, we shall speak of them collectively as boatmen, for the better class of 'company' employees who were engaged in maintenance work could hop on a tug and go off for a couple of days, handling the tug as well as the regular boatman could. Again for the purpose of clarity, we do not propose to distinguish between boatmen and bargees, all men who worked on, in and about inland waterways had many things in common.

Some were dependable:

> We, the underwritten, being the Engineer, Inspector, Foremen, Hagmasters and Workingmen, employed on the works of the Grand Junction Canal, highly sensible of the blessings of our free Constitution and truly loyal to our King, think ourselves called upon, as Englishmen, to stand forth in support of our Laws and Property, and that of our Honourable Employers: and do therefore, associate, under the following Rules and Articles, which we request our worthy Engineer, Mr. Barnes, to transmit in our names, and on our behalf, to the Marquis of Buckingham, Lord Lieutenant of the County of Buckingham, with our loyal request that he will lay it before the King, with our firmest assurances that we will spill the last drop of our blood, in the cause of Old England, against all foreign and Domestic Enemies.[25]

Some were jolly if we are to believe the evidence of a china plate held by a boatman friend of ours:

> Pickford, Beach and Snell's, are jolly Lads & true ones:
> Kenworthy & Worthington's you'll likewise find true blue ones:
> Wakeman, Green & Ames, amiss you'll never find Sirs,
> Holt's Crocket & Salkeld's will sail fast as the wind Sirs,
> True harted & jolly ones you'll find with Heath & Crawley,
> Sturland's Henshall's, Alkins too, can likewise use their mauley,
> So likewise can the Boatmen all, & drink their can of flip, Sirs,
> They'll drink there grog, & toast their lass, & then thay'll
> Crack their whip Sirs.

Some were viewed askance: 'The boater has not the intellectual quickness of the London thief; but his purely animal life, devoid of all spiritual and almost all social influences, has produced in him a low and coarse animalism.'[23] Some had a rough old working life in the bowels of a colliery or tin mine, moving boats by hand from the working face to the transhipment point from whence another boatman would pick it up.

GRAND JUNCTION CANAL CO.

CAUTION TO BOATMEN.

At the Clerkenwell Police Court, on Monday, the 23rd May, 1864, **BENJAMIN WILSON** was convicted of intimidating and compelling the hands to quit the boat "Fleetwood," and sentenced to One Month's Imprisonment, with Hard Labour; and, at the same Court, on Monday, May 30th, **JOHN MUSSON** was brought up on remand, and convicted of assaulting and compelling a Boatman to leave the boat "Grays," and sentenced to Two Months' Imprisonment, with Hard Labour, in the House of Correction.

NOTICE IS HEREBY GIVEN,

That the Hands who joined **WILSON** and **MUSSON** in the above outrages are strictly prohibited from working with the Boats again, and that any Steerer employing or harbouring any of them after this Notice will be instantly discharged.

CITY ROAD BASIN,
2nd June, 1864.

By Order.

Waterlow and Sons, Printers, Carpenters' Hall, London Wall.

Fig. 6.

> The mode of entrance was by means of a bucket in which I was let down to a depth of 70 yards, when we come to a canal, which runs for a considerable distance in a subterranean channel . . . at this canal I got into a barge . . . which was made to advance by the boatman pushing against small staples fastened to the roof . . . in this way I travelled more than ½ mile . . . having returned to the barge I was landed about ½ mile further at another mine.[26]

Yet another class of boatman who had to haul his boat along staples or pegs was the tub boatman employed in Somerset and Devon, where the canals abounded in tunnels but were bereft of leggers.

With the virtual cessation of commercial traffics over the majority of the country's canals, a large number of boatmen have been forced to take jobs on the bank with the various nationalised bodies that have had control over the waterways, or they go into a factory. Some are successful, ending up as a tug driver, with the dredging team or working for a hire-boat company. A very few boatmen have been, or are, on the canal simply because they know no other life. George Harris, who at one time or another has worked for every carrying company, told us one day:

> 'To be quite honest with you we went back boating because we couldn't find nowhere to live, and that was the only way to get somewhere. I think that even now if a good many of these boat-people could get a house they wouldn't be boating, although you get a good many who can't settle down on the land, I mean a boatman don't like regular work, and another thing, if a boatman was to go over to the factory to work they have a foreman over them, which they have never been used to. I mean, they have always been used to their own life and their own tinpot way of going about and they won't be told.'

Some canal men favoured charging up and down one route, the Shroppie perhaps, loading oil at Ellesmere Port and running to the depot at Wolverhampton, a nice easy three days'

29, 30, 31 Canal greeting cards

32 Narrow boats in tow, Manchester Ship Canal, 1957

33 Near Aintree racecourse, Leeds & Liverpool Canal, *c.* 1950

journey. Or they might prefer the coal run from the collieries on the Coventry Canal to Dickinsons Paper Mill and back, ten days for the round trip. Others roamed the canals and if there were any odd jobs, whether it be at Market Harborough, Nottingham, up the Erewash or on to the Leeds & Liverpool Canal, all were grist to their mill. Some boatmen would never go near the Manchester Ship Canal, while others found some pleasure in being towed, five or six pairs behind a tug, across to No. 9 Dock for loading. One boatman's claim was that he had been on every waterway, barring the Droitwich Canal, working at various times for the Severn & Canal Carrying Company operating out of Gloucester, Samuel Barlows out of Braunston on the Grand Union Canal, S. E. Barlow at Glascote on the Coventry Canal, F. B. Lycett & Son from Sneyd Wharf on the Birmingham Canal Navigations, Willow Wren Canal Carrying Company from Hayes Wharf, London, the Anderton Canal Carrying Company from Northwich and Kellogg's from Manchester. In the process he garnered a wife and four children and, to add a little variety to his life, picked up four 'deaders' on his propeller, one at Leighton Buzzard (Beds.), two near Worcester and one at Walsall. He had a brother scalded to death on a barge whilst laid up at the 'Starvation Buoys' on the Thames. He had worked under various systems of payment on a regular wage, on trip money, or a combination of both and the one liked least where the company deducted £5 per week from the wages, theoretically placing it in the bank to cover the provision of and loss of equipment; for the boats were hired out on a bare-bones charter to the boatmen who were responsible for providing their own ropes, cans, chimneys, diesel fuel and even if they broke a shaft they paid for it – 'If your engine blew up you came awfully near to being bankrupt.' As recently as the late 1950s wages for a week's hard work could be only £6, and the next week £28, and then perhaps nothing for three or four weeks.

Canal boating has its own perils and compensations. During the Second World War one pair of boatmen working for Fellows, Morton & Clayton, loaded up with aluminium and were ambling along peacefully on a moonlit night on the Shropshire Union Canal when a German hit-and-run raider dropped a few bombs near enough to shake them. It was not until they reached Birmingham and had their ears pulled that they found out that because they had not clothed up (being under instructions to take the aluminium through at once if not sooner) the aluminium ingots had created a beautiful mirror effect. A number of boats were lost during the bombing of London docks, including the wooden motor boat *Robin* at Limehouse. Fortunately the crews were lurking under a crane at the time.

Boatmen have, among people who know no better, a reputation for being taciturn when going about their occupation and garrulous when in a public house, perhaps the truth lies between the two. One boatman friend of ours, still alive, has probably committed most of the sins in the calendar at one time or another, petty pilfering, absconding with debts, fighting and saying 'rhubarb' to authority, but he had never drunk a drop of alcoholic liquor in his life. 'Boatmen,' said another to us one day, 'are the friendliest people imaginable – amongst themselves!' On the other hand, 'The first salutation a boater meets you with is, "Gaffer, wat's the time o' day? Can you gie us a match to light one's pipe?" and if you oblige them in that or any other little thing, you have won them over, and after this they will be very open and free, and will fight to death for you, if need be.'[23] Some, like most householders, take enormous pride in their home, their boat.

> The spotless neatness of the little cabin, and the last polish bestowed on the brass fittings, are characteristics they frequently have in common with the pleasure yachts of our upper circles. It seems that only on the water can one learn how brilliant a polish brass will take . . . The exterior decoration of these boats is noticeable, and evinces the pride taken in their appearance by the owners, who repaint them with the gayest colours as often as they can afford to do so. On the outside of the cabins are painted two or four landscapes (usually river-scenes) of which they are proud enough . . . The smartness of the cabin part of the barge is often the more striking, from the fact that the load it bears is of a very opposite character, as coal, which is perhaps the most common freight.[27]

Others were perhaps not so particular, either in painting or sanitary habits.

> Their habits are filthy and disgusting beyond conception. I have frequently seen women in a half nude state washing over the sides of the boat as it was moving along, out of the water of the canal, upon the top of which has been floating all manner of filth. They wash their clothes – those that do wash – out of the canal water, and instead of being white, or near to it, they look as if they had been drawn through a mud hole, wrung, and hung out upon the boat line to dry . . . Swearing and drunkenness seem to be the two ruling passions... Swearing is taught to the children before anything else; nearly every sentence begins with an oath, and finishes with blasphemy . . . The women are coarse and vulgar, and, if anything, can outdo the men in resorting to obscene language and disgusting conversation.[23]

However, to put this kind of report in perspective one must not forget the conditions existing in, say, a Leeds slum:

> With broken panes in every window-frame, and filth and vermin in every nook. With the walls unwhitewashed for years, black with the smoke of foul chimneys, without water, with corded bed-stocks for beds, and sacking for bed clothing, with floors unwashed from year to year, without offices . . . while without, there are streets, elevated a foot, sometimes two, above the level of the causeway by the accumulation of years, and stagnant puddles here and there, with their foetid exhalations, and excrementitious deposits on all sides as a consequence, undrained, unpaved, unventilated, uncared for by any authority but the landlord, who weekly collects his miserable rents from his miserable tenants.[28]

The human hand seen under a microscope is one mass of living, heaving, life, no matter whether you use 'Sprightly Snow' or 'Purple Yamac' and there are still a number of houses where neither is used and where bugs are to be found. Even in the 1970s there have been found places where babes were playing on the floor while little white maggots ambled up and down between the floorboards. In the close confines of a cabin such uncalled-for livestock make their proximity terribly uncomfortable and the practice when taking over a 'change-boat', the boatman's regular motor boat or butty boat being under repair, was to ensure that it had been fumigated.

In the early 1870s there were a number of 'do-gooders' who found that their presence in the emergent nations was unpopular and so they turned their attention to canal people. One found it astonishing to hear 'You must look for yourself, or you will get pitched into the cut' when he asked such questions as, 'When was your boat bugged and cleaned?', 'Do you tie up on Sundays?', 'Can you read and write?', 'How is it you stare so?' and 'Why is your horse so thin and can scarcely walk along?' Some of the nitwits who went around making inquiries – their descendants are still employed in the nationalised industries today!– were fired with a religious belief. Edgar Wallace in his Sanders of the River novels castigated the missionaries who went out to teach the heathen about Christianity, omitting such fundamental requirements as a knowledge of the native language and the local customs. They preached to boat people in the same manner and having an axe to grind they chose the least salubrious craft and crews to question. In 1874 one such made the profound statement that 'For many years the name of "Bargee" has been synonymous with ignorance, coarse vulgarity, and even murderous brutality', and, appealing to Victorian virtue, his conclusion was that

> We shall, as a nation, have something to answer for some day for neglecting to protect the boatwomen and children. We seem to forget that in taking care of the children we are also providing for our own safety. We seem to forget that in taking care of the children we are giving physical stamina to future generations, instead of propagating a race of short, stunted, decrepit, pale, sickly men and women, who become aged at 50 . . . Let us take timely warning, and rescue these miserable women and children from the physical degradation, moral pollution, and unhealthy atmosphere they have to endure in those hot, damp, close, stuffy, buggy, filthy and stinking holes called boat cabins.[23]

By 1879 matters had obviously reached a climax:

> The boat system of inland navigation has gone on increasing year by year, until at the present time it presents one complete network running through the length and breadth of the land, from the centre to the circumference, of something like 4,710 miles in length. On its surface are over 25,000 boats, carrying human beings of all ages, together with filth, mud, manure, and the refuse of our large towns. Some of the cabins are models of neatness, and a man and two youths might pass a few nights in such very comfortably. Others are the most filthy holes imaginable; what with bugs and other vermin creeping up the sides, stinking mud finding its way through the old leaky joints at the end to the bottom of the cabin, and being heated by a hot stove, stenches arise therefrom enough to make a dog sick. In the boat cabins – 'hell holes', as some of the women call them – people of all ages live day and night. Fathers, mothers, sisters and brothers, sleep in the same bed and at the same time. In these places girls of 17 give birth to children, the fathers of which are members of their own family . . . It is really sickening and disgusting to witness habits indulged in by these useful people that animals would pass by.[23]

It is true that the majority of boatmen could not read, but they would become aware of what was being written as things could be read to them. If you went to the inhabitants of a town with a population of 25,000 say, Bromsgrove in Worcestershire, Fleet in Hampshire, Gainsborough in Lincolnshire, Loughton in Essex or Inverness and you told them that their houses were the most filthy holes, they had bugs, they committed incest and were lower than animals, you might possibly find a few of whom this would be true but how would the other 24,950 feel? Might not they grow a little taciturn?

A common practice that no boatman liked was for a minister of the Church to stalk on to the wharf, unsmiling and without a pleasant word, and step straight on to a boat intent upon asking silly questions. A far better approach was through the 'Boatmen's Chapels', set up at most of the terminal points of waterways. Basically places of worship, they made a pleasant change from the nearby pub and one could always be assured, especially where the Salvation Army were concerned, of a pleasant word, a drink, sometimes even a meal and a game of dominoes. Whether these chapels were successful in their early days – most were founded after 1860 – is a matter of opinion. One minister-writer berated them, saying that with 60 boats tied up in the basin on a Sunday and only two or three families going to the chapel they were a total failure. On the other hand there are many extant letters written by, or for, boatmen in appreciation of the kindnesses they received. The upper middle class may have been, in the Victorian era, the backbone of England but unfortunately they had a tendency to regard those 'of the lower orders' as 'not being quite the same as us', and some of their proposals must have been the most unsympathetic that could have been dreamt up. 'Could not little singing parties be made up out of our elder scholars and teachers to visit canal wharves, back slums, and prisons . . . capable and willing to talk to the boaters and others in language they can understand . . . which would, in my opinion, go further towards drawing the boaters and "jail birds" to the church.'[23] One can imagine a boatman, having worked through the night, whacked out, sleeping in his cabin being awakened by a 'little singing party'. No doubt his attitude would confirm that which they already presumed to know. Queen Victoria, by contrast, showed a more practical approach. 'At the commencement of this year a poor boatwoman, with a husband and three little children, had an addition of three other little babies added to their stock in the cabin. Her Majesty the Queen, who is always spoken of by the boaters as one of the best women in England, kindly sent the usual bounty of £3 given on such occasions.'[23]

Needless to say, over the years boatmen have developed a dialect of their own, although even this varies from region to region. Some words are relics of the early nineteenth century, others debasements of common words. 'Gullontyne Lock' for guillotine lock, 'I dain't gew', 'Hackerdock' for aqueduct, are among the latter. If a pleasure craft steerer is told by a boat's captain to 'hold in' he should move towards the towpath, for waterways do not have port and starboard as at sea, but an inside and an outside. 'Check yer moter' means to slow down, derived from horse-drawn days when the horse, by checking, would take the way off the boat. A corner or bend is always 'a turn', if slight it is 'a little turn', if severe 'a big turn'. To a canal boatman all weathers are much of a muchness, except the east wind and that is 'a lazy wind, it gews straight through yer, cos it carsn't be bothered to go round yer'. Locks, wharves and other appurtenances of a waterway have, over the years, attained different names to those shown in official records, the best-known being Sutton Stop for Hawkesbury Junction, named after a family of lock-keepers who lived there. To a boatman Flowers Lock is better known than the Top Lock, Bascote of officialdom. 'Footies Wharf' is another name for Tibberton Wharf on the Worcester & Birmingham Canal, the family after whom it is named claims to have had no less than six generations living simultaneously in boats on the waterway at one time. 'Maffers' for Marsworth is probably a laziness, but the 'top of 'ampton' is the best way of describing the top of the flight of 21 locks at Wolverhampton. If traversing the main line of the Grand Union a boatman would refer to the 'Nag's Head Three Rise' (locks nos 34–6) named after the local hostelry. Number 50 on the other hand would be Bush or Barkers lock, whilst number 52 could be called 'The Crooked Billet' or 'Awkward Bills' or 'Old Neds' – the latter being derived from Ned Adlum who worked as a navvy when the cut was first dug. 'Sweeps' commemorated Eli Oliffe, once a chimney sweep, who later kept the famous 'General Boatmens Stores' alongside lock 55, Berkhamstead. The three locks at Winkwell – Irishmans – gained their nomenclature following the death by drowning of a lock-keeper of Irish origin. Not only were lock-keepers and pubs commemorated but at number 80 we have heard the names, Walker's, Beasley's and Cherry's, three one-time Canal Company overseers who had previously lived there. Denham Deep (No. 87) is the deepest lock on the Grand Union, whilst Widewater (No. 86) is a logical self-explanatory name.

As the true boatmen fade from the scene (most of them are lying in Braunston churchyard) so a new generation is coming to the water and with a new type of boat. Some have picked up the appellation of 'noddy boatmen' from the fancied resemblance of their craft to jelly-moulds and their steerers to a television personality; others start off with a small cruiser and dream of the days when the canals were canals, viable, living entities, and it was possible to work a boat making a living in the process.

Twas up in Brummagem so I do hear say,
A boat by the name of the *Rosemary* lay,
She was clothed up and painted in traditional style,
But she hadn't carried for a very long while.

Along came a boatman the old boat to see,
Says he here's a craft that is useful to me,
'I'll load her with coal and to London I'll steer'
said the boatman to the owner, 'if I take her from here'.

The owner said 'Yes' the boatman 'Okay',
And into the cabin he went straightaway,
He lit up the stove, cleaned cobwebs and mould,
And he polished the beam till it shone like fine gold.

He sang as he laboured far into the night,
Got up in the morning before it was light,
He put the rusty blowlamp on the cylinder head,
''Tis a fine day for boating' the old boatman said.

He primed up the engine, a prayer in his heart,
And kicked on the flywheel to see if she'd start,
With a bang like the noise of a ten-pounder gun
The aged Bolinder started to run.

He cast off the fore-end at the counter he stood,
As *Rosemary* shook herself free of the mud,
With tears in his eyes says the boatman, 'We may
Get right down to Coventry for the end of the day.'

On dark stormy nights round the fall of the year,
If the beat of a Bolinder distant you hear,
It's not Claytons *Stour*, *Youmea* or *Tay*,
It's the ghost of that boatman and the old *Rosemary*.[16]

6 Goods in Transit

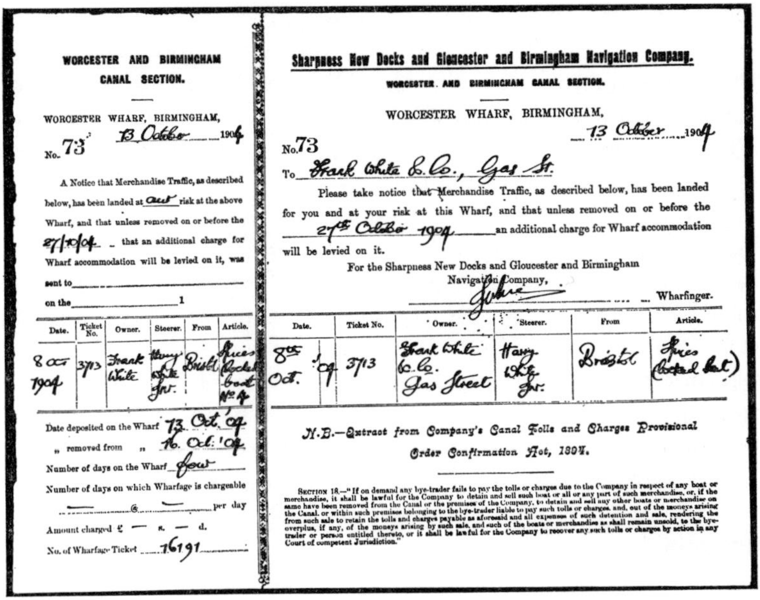

WORCESTER AND BIRMINGHAM CANAL SECTION.

WORCESTER WHARF, BIRMINGHAM, 13 October 1904

No. 73

A Notice that Merchandise Traffic, as described below, has been landed at our risk at the above Wharf, and that unless removed on or before the 27/10/04 that an additional charge for Wharf accommodation will be levied on it, was sent to ______ on the ______ 1

Date.	Ticket No.	Owner.	Steerer.	From	Article.
8 Oct 1904	3713	Frank White	Harry White Jnr.	Bristol	Spices

Date deposited on the Wharf 13. Oct. '04

„ removed from „ 16. Oct. '04

Number of days on the Wharf four

Number of days on which Wharfage is chargeable — @ — per day

Amount charged £ — s. — d.

No. of Wharfage Ticket 16191

Sharpness New Docks and Gloucester and Birmingham Navigation Company.

WORCESTER AND BIRMINGHAM CANAL SECTION.

WORCESTER WHARF, BIRMINGHAM, 13 October 1904

No. 73

To Frank White & Co., Gas St.

Please take notice that Merchandise Traffic, as described below, has been landed for you and at your risk at this Wharf, and that unless removed on or before the 27th October 1904 an additional charge for Wharf accommodation will be levied on it.

For the Sharpness New Docks and Gloucester and Birmingham Navigation Company,

______ Wharfinger.

Date.	Ticket No.	Owner.	Steerer.	From	Article.
8th Oct. 04	3713	Frank White & Co. Gas Street	Harry White Jnr.	Bristol	Spices

N.B.—Extract from Company's Canal Tolls and Charges Provisional Order Confirmation Act, 1894.

SECTION 18.—"If on demand any bye-trader fails to pay the tolls or charges due to the Company in respect of any boat or merchandise, it shall be lawful for the Company to detain and sell such boat or all or any part of such merchandise, or, if the same have been removed from the Canal or the premises of the Company, to detain and sell any other boats or merchandise on the Canal, or within such premises belonging to the bye-trader liable to pay such tolls or charges, and, out of the moneys arising from such sale to retain the tolls and charges payable as aforesaid and all expenses of such detention and sale, rendering the overplus, if any, of the moneys arising by such sale, and such of the boats or merchandise as shall remain unsold, to the bye-trader or person entitled thereto, or it shall be lawful for the Company to recover any such tolls or charges by action in any Court of competent Jurisdiction."

Fig. 7. Warehouse Weighbill.

One of the complaints of all railway traffic managers was to the effect that 'traders are wont to leave their goods in our wagons, on our sidings, for an excessive period; in effect providing cheap warehousing'. This difficulty was also well known to the owners of day boats, for the boat might lie three or four days with no demurrage paid before the recipient was prepared to discharge the cargo of coal or slack. Long-distance carriers were better organised and should there be any difficulty in unloading, say, because of Wakes Week, the goods could be off-loaded into a warehouse and transmitted onwards at a later date. As the cost of labour – until the late 1930s – was negligible and rates minimal this satisfied everyone. But should a small carrier be forced to leave his goods at another company's warehouse the story was different, sufficiently so for polite but useful forms to be printed, as the one reproduced above. In this case the steerer, Harry White, had had a leak in his boat and rapidly off-loaded the spices into the old Worcester & Birmingham Canal transhipment warehouse, doubtless intending to collect after repairs had been done and continue on his way. In this case a fortnight's grace had been given; at Preston Brook a 48-hour allowance was common practice. On the Duke of Bridgewater's canal, traffic and hence warehouse congestion was such that boats had often to wait for days to unload, and were then charged for the time they had occupied the basin!

One cannot but be astounded at the complacency and overbearing attitude of a wealthy company when it is reported that in 1828: 'Notwithstanding the number of tons of flints and stone unladen weekly on the Duke's quays, and the present improved state of machinery, the primitive process of weighing by hand has not as yet been replaced by a better. Every individual stone passed through an ordinary scale. The machine is placed on the quay, and the stones are slung out of the vessel by a rough sort of crane.'[29] Often too, the canal companies were very retarded in providing wharfage facilities when a traffic developed.

Loading wharves were scattered up and down the Coventry and Oxford Canals, all in use until the late 1920s, and boats might have to call at one specific wharf or have to hunt for a load. On the Coventry the first stone quarry was at Mancetter where loading was by means of a shute, the location being known to boatmen as 'Tipitsy Wide 'ole'. After that there were the Able Stone Quarry, loading again by means of a shute, and a quarter of a mile on, at the Anchor Bridge, the Gees Stone Quarry relied on loading overside. This meant shovelling the chippings into the bottom of the boat, 20 tons at a time. A two-mile run then to Spring Wood and Boon's Stone Quarry near Nuneaton. A mile on you came to the Punch Bowl where refreshments would be taken before inhaling that inexorably foul dust that came down with the chippings from the shute at Judkin's Stone Quarry. If you were not loading there you went on through the bridge to the 'Black 'ole', a dark blue granite quarry owned by the Midland Granite Quarry Company. Even if you could not get a load there you still had one more chance at the Griff Granite Quarry a short way on. Sometimes instead of taking this granite down the Oxford Canal you would turn right at Braunston on to the Warwick & Napton Canal Navigation and then you paid tolls at Wigrams Turn (Calcutt). The permit you garnered there enabled you to proceed to Warwick Junction and on to Warwick Basin where the only means of discharging was by hand. Boatmen reckoned to get 20 tons of chippings out of a boat in three hours. If you were going on past Warwick to the Rising Brook Bridge and along the line of the Warwick & Birmingham Canal Navigation there was another gauging office at the bottom lock of Hatton, there you were gauged again and issued with another permit which would take you to Camp Hill. All this wasted time and effort was a matter of course and singularly unprofitable to both canal company and boatman.

One displeasure often found was the variation in the density of water which could materially alter the apparent load on a craft. To quote a boatman: 'With this boating I reckon you carry a lot of tonnage for nothing. I mean you go down to a place like Brentford and you load up a load of deal, you can mark your boat with the tonnage that's on it and then you go to a place like Pooley, back of Coventry. You can bet your boats deeper at Pooley than it was with the proper weight on, because of the water you carry more for nothing there than what you do anywhere else.' Obviously there were many disputes over the correct tonnage in a craft and, fiddles disregarded, the toll collector was the ultimate power in the land.

> The engineer will also have to make provision, while the canal is digging, for stop-bars at the several intended toll-houses, or other places where it may be necessary to stop barges in the night, or in the case of any dispute about their lading: these bars are composed of a large baulk of fir timber floating on the water; and a small arch capable of containing such a floating beam of the proper length is provided under the bank, so that when the trade on the canal is required to be stopped, the toll-clerk has only to draw out the beam by means of a cord attached to it, until its end enters a recess in the opposite wall, and then to lock the beam fast.[30]

Naturally, most toll clerks were unwilling to undertake such a final move. Visualise, if you will, a dark and gloomy night somewhere on the Birmingham Main Line, eight o'clock of a winter's evening, the toll clerk makes up his books, a thump on the door and he goes out to see the boat captain whose craft he has just stopped for the sake of a ton, and who knows what might happen? 'During last year nearly eighty boaters were confined in Stafford Gaol – nearly all of them young men under thirty years of age – for various offences, principally thieving; and at the present time there are eighteen undergoing sentences in the gaol.'[23] It was not in any way uncommon for a toll-keeper, whose house was sufficiently remote, to be attacked when carrying home the dues he had received – fortunate were they who lived in a toll-house on the island in the middle of the canal. Others occasionally embezzled the loot, a singularly unprofitable occupation for which they almost automatically got seven years' transportation if caught.

Reverting now to wharves – it must be understood that not all wharves had out-of-date equipment, the bigger companies might have cranes, howbeit of the manumatic pattern, but geared so that one man could, with relative ease, lift five tons at a time. At a mill, granary or timber yard there might well be covered-in accommodation, rather akin to a railway station canopy, whereunder the boat could lie protected from the elements. In turn there would be an overhead hoist, often steam operated. Covered accommodation was essential for some perishable items which tended to be packed in sack-bags and could not always be put in the main warehouse immediately. Particularly noticeable in this field is Cadbury's old 'Riverside' Works, at Bournville, where four boats could lie simultaneously under cover for the handling of cocoa beans or chocolate crumb. Flour was yet another 'protected' load, which all too often had to be 'humped' on board, one sack (of 120 lb.) at a time on the boatman's back. Due to its characteristic of becoming a hard inert mass – 'lumping' – this was almost a more unpopular exercise than loading sacks of wheat. A number of boatmen are never seen without shirts, having picked up a form of dermatitis on their backs from carrying wheat, the chaff getting ground into the skin. Brown, or demerara, sugar was another 'humped' traffic, and one which called for the most careful clothing-up (covering with a tarpaulin) on the boat. Some works would, if lacking in a canopy, decline to unload the craft in inclement conditions. Chests of tea also required this kind of coddling, while loose wheat is known for its flourishing growth in the humid conditions of a poorly covered-in boat. Bulk salt had its own tricks; often carried in open boats and merely shovelled out on to an open wharf the wastage must have been unnecessarily high. Apropos of some of the more valuable bagged loads, one essential adjunct to the trade was a good standard of boatmen, and it is known that at least one very good and profitable run, lost ostensibly due to a change of transport pattern, was in fact withdrawn due to the uncontrolled, unhygienic habits of some boatmen's children and animals.

A few goods, specifically timber and wheat, were (some long time before the Americans thought of the word) palletised; the timber being cased, while 20 or 30 tons of wheat were laid into the boat inside a wooden framework, strengthened and equipped with eyes for lifting by crane. Steel bars, rods or strip were commonly bundled, being held together with wire or steel bands which frequently came adrift when being craned overside from ships in the docks. If they survived that not a few went overboard at the other end! Some wharves are very scarred indicating that they have been utilised for rubbish and scrap, the former being tipped directly on board if a boat were ready, if not a dumper truck would scoop it on board. Discharge, once by shovel, is now mechanised using a crane and grab. Scrap, profitable but loathed, inevitably comes away from its bands, and has to be loaded on, bit by bit, by hand. Mouldering rusting heaps on obscure city wharves often denote the point at which a cheesed-off boat captain 'jacked it in' and went home.

In later years modern factories brought electric hoists into service and either a continuous Archimedean screw or a conveyor belt had some specialised usage.

Looking through a list of wharves, produced by Fellows, Morton & Clayton in 1918, the concentration of them within a few miles of Birmingham is proof enough of traffics. On the London run we have firstly the Birmingham Bar Stop Lock and

Toll House; two furlongs away, at Fazeley Street, the Warwick Public Wharves; another 400 yards to the warehouses, wharves and offices of Fellows, Morton & Clayton and, virtually next door, one lock up on the Bordesley flight of locks, the City of Birmingham Gas Works, Adderley Street. At the Top Lock are the public wharves at Sampson Road and Bordesley Toll Office. On the ten-mile pound from thence to the Top Lock of Knowle we come to wharves with famous names, Jones, Manley & Co., timber merchants, the Birmingham Small Arms Company, toolmakers, City of Birmingham Refuse Disposal Depot, Alldays & Onions, engineers (and *inter alia* makers of quite fine motor cars), Symmonds Bros., the City of Birmingham Stone Depot, the not very appositely named Golden Hillock Road and Birmingham Small Arms again at Small Heath. At Hay Mills we find Pontifex, engineers, Whitfield, iron-founders, and Derringtons, brickworks. At Yardley there was the coal wharf of the once famous carrier T. Boston, then Yardley tunnel wharf, an almost worked out gravel-pit and three coal wharves,

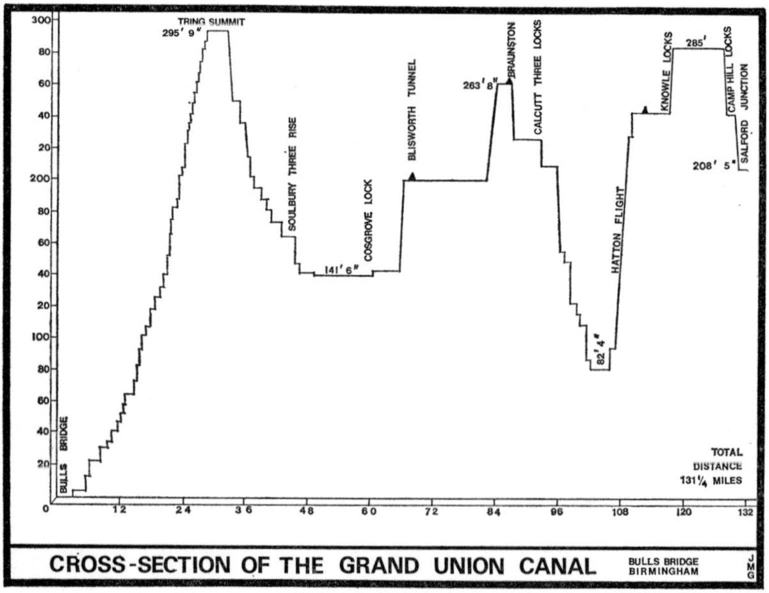

CROSS-SECTION OF THE GRAND UNION CANAL

Fig. 8.

Yardley Road, Francis Road and Lincoln Road. Continuing along the same cut we have public wharves at Olton and Solihull, together with the Solihull Gas Company's own coke and coal depot. Next is the complicatedly named Catherine de Barnes Heath Wharf – abbreviated by boatmen to Catley, or Ketley, Barnes. If by this time, still only eight miles out from Birmingham you have a thirst and have missed the Anchor at Solihull, the Boat at Ketley Barnes will restore life. Another five wharves occupy the next four miles from the Bottom Lock of Knowle and at Heronfield three overnight accommodations are shown (for horses, not men), the Kings Arms, Ivy House and the Cat-in-the-Window. Three furlongs on the Old Black Boy offered similar accommodation, then on round to Hatton Toll Office and you pass another eight wharves – a total of 36 in the 20 miles from Birmingham, and Hatton Asylum half a mile away has yet another!

The Warwick & Napton Canal from Budbrooke Junction, where it met the Warwick & Birmingham Canal, through to its junction with the Oxford Canal at Napton, offered a plethora of wharves, all too many, alas, now disappeared. Your weighbill might direct you to Warwick Locks, Cape Locks or Bayleys' Two Locks – but all are one and the same. The next wharf, not inappositely was the Warwick Union Warehouse! Guy's Cliff, Nelson Dale, the Bakehouse or the Mill at Emscote might all call for your services; coal, coal and more coal. Next came the Leamington & Warwick Electric Power Station, the Corporation Tip, Leamington Gas Works and the Midland Electric Light Company. Nelson & Co needed slack at their oil mills, whilst Sidney Flavel used both iron and coal to manufacture his range of gas- and coal-fired stoves. The fifth wharf from there – five miles from the Junction, was the Radford Old Brewery and very welcome it was!

In the thick of the locks, the six-rise of Radford, the four of Bascote and the ten of Stockton, nine wharves are interspersed, including the Cement Works, one of the last traffics that was operated under the loving hand of the British Waterways Board. Another five wharves and the Toll House at the top of Wigrams three-rise (Calcutt) was in sight and another 14 miles were completed.

To give some idea of the work involved at a toll-house it will suffice to say that in 1925 no less than 117,000 tons were handled at Stewponey, giving a revenue in excess of £4,000 all collected in dribs and drabs.[31] Typical tolls which the collector might have to extricate, not necessarily in cash, for many would be debited to either the Carrying Company or the consignor, are shown here, drawn from the toll book at Windmill End, Birmingham Canal Navigations. These tolls appear to be those applicable from 1908 to the railway groupings of 1923, remaining, of course, constant throughout the period.

	From	*To (via)*	*Distance (miles)*	*Article*	*Toll per ton in old pence*	*Special toll per ton*
1.	Atlas Tube Works	Withymoor Station	½	tubes	0·60	
2.	,,	Bumblehole Tip	1¼	mud	min. 1*s*. 6*d*.	plus 1*d*. per ton
3.	Bumblehole Branch Top	Selly Oak Wharf	8¾	slag	6·875	
4.	,,	Worcester Canal	8¾	slag	5·375	
5.	Bishton Bridge	Coombs Wood Colliery	3½	sand	3·975	plus 125% extended voyage
6.	,,	Webbs Green, Halesowen	5	manure	4·50	
7.	Coombs Wood Tube Works	Autherley	13¼	tubes and spelter	11·625	thro' traffic
8.	,,	Berkeley Street	13½	spelter	16·35	
9.	,,	,,	13½	scrap brass	18·65	

34 British Waterways Board, Wakefield Depot

35 Leeds Co-operative Society tug *The President* with two barges, Leeds Lock, 1970

36, 37 Boat building at Ashvale, Basingstoke Canal, in the 1920s

	From	*To (via)*	*Distance (miles)*	*Article*	*Toll per ton in old pence*	*Special toll per ton*
10.	Coombs Wood Tube Works	Canal Depot Langley, Titford Canal via Bradeshall and Oldbury Locks	12¼	spelter	15·4625	
11.	„	Stour Canal via Blowers Green and Black Delph	10¼	iron	9·7625	
12.	„	„	„	tubes	6·125	Gloucester traffic
13.	Dudley Wood Lime Kilns	Webbs Green Farm	5¼	lime	4·625	
14.	Fly Colliery Basin	Stour Canal	5¼	coal	7·35	
15.	„	„	„	„	6·00	B. Pearson & Co. A/c
16.	„	Langley Forge via Netherton Tunnel		„	14·50	W.H.Matty & C. A/c
17.	Gawn Colliery	Minworth Top Pond via Netherton Tunnel and Farmers Bridge	22	slag	13·75	
18.	Gorsty Hill Co. Wharf	Stour Canal via Two Lock Line Black Delph	5¾	coal	9·55	

	From	*To (via)*	*Distance (miles)*	*Article*	*Toll per ton in old pence*	*Special toll per ton*
19.	Gawn Colliery Basin	Whitton Turnover Bridge via Netherton Tunnel, Toll End and Perry Barr	14¼	blast furnace slag	12·8125	T&S Element A/c
20.	,,	Boat and Slipper Wharf via Netherton Tunnel and Smethwick	7½	slag	9·25	
21.	Halesowen Wharf	Stour Canal (Gloucester) via Black Delph	11¼	iron	9·625	1*d.* in addition if crane is used
22.	Hawn Basin	Coalite Works Wednesfield, via Netherton Tunnel, Factory Locks, Horseley Fields Junction	15¾	bricks	16.7875	
23.	Knowle Colliery	Pensnett Canal via Park Head Locks	4¼	iron stone	4·5375	
24.	,,	California Brick Works via Netherton Tunnel and Bar Lock		coal	9·00	
25.	Lodge Farm Canal Company's Basin	Waterfall Lane	3	timber	6·60	

	From	To (via)	Distance (miles)	Article	Toll per ton in old pence	Special toll per ton
26.	Marine Works	Rowley Gas Works	2½	rubbish	3·00	
27.	Netherton Iron Works	Knowle Colliery Basin entrance to off side	1¼	dredgings	2·75	If emptied into canal 5*s*. per boat in addition
28.	,,	Gorsty Hill Wharf		coal	2·85	Exclusive of wharfage
29.	Old Hill District	Stour Canal *for Kidderminster only*		bricks		BCN apportionment 6*d*. special
30.	Pennant Hill Colliery	Spon Lane Wharf via Netherton Tunnel and Spon Lane	5¾	coal	13·55	
31.	Rowley Pottery	Warwick Canal for London via Netherton Tunnel and Farmers Bridge	11¾	pipes and sinks	10 ·875	through traffic
32.	Springfield Basin	Bell Pond via Netherton Tunnel and Toll End	11	undressed road stone for repair of roads	11·00	
33.	,,	Winsom Green Workhouse	9	kerbs and setts	12·1	
34.	Two Locks Top Lock	Rowley Stop Ovens	2¾	firebricks	4·30	
35.	Windmill End	Primrose Hill Wharf	1¼	slag	Minimum 1*s*. 6*d*. and 1·00*d*. per ton	

	From	To (via)	Distance (miles)	Article	Toll per ton in old pence	Special toll per ton
36.	Warrens Hall Colliery	The Pensnett Canal via Netherton and Park Head		coal	2·85	stoppage at Primrose

All the above figures are pre-decimalisation. There were 1,455 entries in the book but certain factors are evident from these few entries. For example, from Bumblehole to Selly Oak Wharf, both points on the Dudley Canal (3) is 1·5*d.* higher than that for traffic of an identical distance to the Worcester & Birmingham Canal (4); this was the drawback previously referred to. The toll from Bishton Bridge to Coombeswood Colliery (5) allowed for transhipment *en route*. Coombeswood Depot Works (7) to Autherley represents an agreed rate for traffic passing onto the Shropshire Union Canal. (8) and (9) shows the differential between two different traffics, being a case of what the traffic would stand. Again (11) and (12) show a different rate applicable for local and through traffic. Traffics working from the Fly Colliery Basin led to quite a few upsets (14) and (15), the preferential rate was only legal if there was sufficient guaranteed tonnage to keep tolls at an overall low level; carriers paying the higher rate were prone to kick up a fuss when such preference was known. The cranage toll of 1*d.* (21) was per ton, and was designed to cover the cost of labour operating these manumatic devices. (10), (17), (21), (22) and (31) show relatively long-distance workings. (27) sounds rather odd but probably allowed for infilling of an unwanted arm. Birmingham Canal Navigations apportionments had been granted over various routes for traffics originating and terminating on other canals, alternatively the loads could be for use by the engineers of the Birmingham Canal Navigations for which special rates were granted (29). The differential between (32) and (33) is interesting, showing again the taking of the traffic for what it would stand. One of the advantages with the Birmingham Canal Navigations was that if one waterway was stopped up there was nearly always an alternative (36). Gorsty Hill (18) is the original spelling, taken from the toll register, the 'r' being dropped these days. Significantly a number of the waterways mentioned are no longer navigable, including (10), (18), (22), (23) and (32).

The different type of goods carried and the varying destinations might not at first sight be surprising on the Birmingham Canal Navigations, but there was little difference even on straight up and down canals. A toll register for the Worcester & Birmingham Canal during a quiet month, February, at a time when traffic was already in decline (1924) shows that 38 boats were gauged, having an average load of $28\frac{1}{4}$ tons. Some boats already had their toll tickets and entries were merely made to cover the return working.

From	*To*	*With*	*Destination*	*To load*	*No. of craft*
Gloucester	Old Union Mills	Grain	Cannock Chase Collieries	Coal	1
"	Severn Co's Wharf	Sugar	Gloucester	Empty	1
Diglis	Birmingham	Empty	Lowesmoor Wharf	Coal	2
Gloucester	Old Union Mills	Wheat	Gloucester	Empty	2
"	"	"	Framilode	Coal	1
"	"	"	Conduit Colliery	"	1
Tewkesbury	Grahames Wharf	Flour	Tewkesbury	Empty	3
Gloucester	Severn Company's Wharf	"	Gloucester	"	1
"	"	"	"	Sauce	1
"	Central Mill	Wheat	"	Empty	2

From	*To*	*With*	*Destination*	*To load*	*No. of craft*
Gloucester	Old Union Mills	Wheat	Cannock Chase Colliery	Coal	2
,,	Coventry	Maize	—	—	1
,,	Severn Company's Wharf	Chocolate	—	—	1
,,	Coventry	Maize	Worcester	Coal	1
,,	Chester Street	Deals	Cannock Chase Colliery	,,	1
Worcester	Birmingham Canal	Elm Boards	Lowesmoor	,,	1
Gloucester	Tipton	Carbide	Gloucester	Empty	2
Not specified – through ticket			Hardwicke	Coal	2
,,	,,		Gloucester	Empty	2
Gloucester	Severn Company's Wharf	Sundries	,,	,,	2
,,	Coventry	Barley	,,	,,	2
,,	Cape Arm	Sawn Wood	Cannock Chase Collieries	Coals	1
,,	Midland Mill	Grain	Glascote	Coal	1
Not specified – through ticket			Frampton	,,	1
,,	,,		Diglis	,,	1
Gloucester	Old Union Mills	Wheat	Bournville	Slack	1
Not specified – through ticket			Cannock Chase Collieries	Coal	1
Gloucester	Chester Street	Deals	Whitminster	,,	1

In the period pre-1850, as post-1930, most traffic was in the hands of large carriers and competition was fierce. From Nottingham in 1832 Thomas Pickford & Co. and Robert Marshalls combined to offer services to London, Liverpool, Manchester, Derby, Bath, Bristol, Hull, Birmingham and 'nearly all parts of England by fly boats, daily'. If, however, you wanted to send your goods to Derby, Grantham, Liverpool and all intermediate places, Samuel Thorpe would be pleased to accommodate you. Robert Barrows offered a service every Tuesday and Friday night to Liverpool, Manchester, Newark, Gainsborough and all parts of the west of England and Cheshire, while German Wheatcroft & Sons went similarly but 'several times a week'. Rather improbably, John Simpson advertised that his boats went to Gainsborough, Hull, Cromford, Mansfield, High Peak Railway, Retford, Liverpool and all parts of Scotland and the west of England. Deacon Harrison's fly boats, on the other hand, found sufficient traffic to run solely to London every Monday and Friday.

In *Aris's Birmingham Gazette* on 9 November 1807, Mr Pickford advertised 'expeditious canal conveyance by fly boats on the line of the Coventry, Oxford and Grand Junction'. On 8 February 1808 the Wolverhampton Boat Company, operating from that place to London, offered 'expeditious canal carriage' and on 2 April 1810 we find William Judd & Sons offering to carry your goods by water from Birmingham to Sheffield or Derby. Some carriers, not unnaturally, favoured certain routes – where there were alternatives – and put forward incentives why they should both carry and use these routes. Thus out of Birmingham in 1820 it was claimed:

> Boats from the wharfs within the town; Birds, Whitehouse's, Robinsons and Crowleys, are capable of delivering goods in London one whole day sooner by the latter route [the Stratford] than they can do by the other [the Warwick & Birmingham] and the merchants and ironmongers in the metropolis are hereby informed of that circumstance. The boat owners by proceeding on this route are necessitated to advance a small sum of immediate money, for tonnage, more than they do on the other route, to counterbalance that, the boats are

CROWLEY, HICKLIN, BATTY, AND Co.

Carriers

BY FLY BOATS

DAILY TO AND FROM

LONDON, LIVERPOOL, MANCHESTER,

SHARDLOW, GAINSBOROUGH, DERBY,

STOURPORT, WORCESTER, BRISTOL, BANBURY,

OXFORD,

AND ALL INTERMEDIATE PLACES.

For Particulars apply to their Establishments,

London,	Salisbury Arms, Cow Lane, West Smithfield.
....	City Road Basin, Albion Wharf, No. 30 and 31.
Wolverhampton,	Union and Commercial Wharfs.
Birmingham,	Crescent Wharf.
Shrewsbury,	London Warehouse, near Welch Bridge.
Worcester,	Commercial Wharf, Lowesmore.
....	Company's Wharf, Diglis.
Banbury,	Company's Wharf, late Judd's.
Oxford,	Company's Wharf, late Judd's.
Liverpool,	Duke's Dock. R. Dickenson, Agent.
Manchester,	Castle Field, Thomas Woodcock, Agent.
Preston Brook,	H. Bayley, Agent.

Crowley and Co. recommend their Friends to forward all Goods intended for them, to the care of the following Public Wharfingers and Carriers.

From ***Paddington,***	To Hill's Wharf, No. 7.
Bristol,	For *Worcester,* per Soule's Trows.
Bristol,	For *Stourport,* per John Jones's Trows.
Gloucester,	To J. D. Walker's Warehouse.
Tewkesbury,	To John Jeyne's Warehouse.
Stourport,	To John Jones's Wharf.
Hull,	Per Richard Furley, Office, 28, Bishop Lane.
Gainsborough,	To Richard Furley's Wharf.
Shardlow,	To Sutton and Co.'s Wharf.
Sheffield.	Per S. Lyster, or Gamble and Co. to Derby,
Dudley & Tipton,	To Whitehouse and Sons.
Derby,	To Sutton and Co.'s Wharf.
Leamington,	W. Maxey's Wharf.
Warwick,	W. Masters's Wharf.

Lock up Boats for the safe Conveyance of Wines and Spirits.

Goods requiring particular despatch would be expedited by being sent to Albion Wharf No. 30, and 31. City Road Basin, Regent's Canal.

Fig. 9.

exempt from the wear and tear of passing through twelve locks, and an extra day's expense; therefore when both circumstances are taken into consideration, the expenses cannot vary much either way, and to the London merchant one day is, at times, of the utmost importance.[32]

In 1860, when the railroads already had a stranglehold, a few of the carriers in Cheshire were pleased to advise of their existence. From Chester boats went daily to Liverpool, Manchester tand Ellesmere Port. William Clowes, wharfinger, would be pleased to be of assistance at Wardle. At Audlum the Shropshire Union Railway & Canal Company found it sufficient to state that they operated to all parts, but Mr John Jones, the local wharfinger, apart from handling the SUR & CC's boats was pleased to place orders for transport with Mellor, Castle & Co., daily to London. At Congleton we had the Bridgewater Trustees offering a daily service to Manchester, Liverpool and the Potteries, whilst the Manchester, Sheffield and Lincolnshire Railway Company went 'to all parts by canal and rails'. At Middlewich it was sufficient to state, 'The North Staffordshire Co's Canal [The Trent & Mersey, or Grand Trunk] passes through the town. At the wharf near the principal street commodious warehouses have been erected for the reception and transmission of merchandize. Mr. Thomas Ball, Wharfinger.'[21] Also represented at Middlewich were the Bridgewater Trustees, Grand Junction Canal Company, Shropshire Union Railway & Canal Company and James Cockshott. Ellesmere Port, Macclesfield, Stockport, Sale, Altrincham, Lymm and Runcorn all had their share of carriers. There were services from Dukinfield: 'Boats leave the Wharf Inn, Wharf Street, for Hyde and Marple, daily at 10.40 a.m. and 3.30 and 6 p.m.'; from Hyde: 'Swift Packets leave the Navigation Inn daily for Manchester at 11.40 a.m., 4.20 and 6.35 p.m.' and Runcorn: 'The Countess Grosvenor Packet Boat leaves the top locks, Monday, Wednesday and Friday for Manchester calling at the intermediate stations', these services being run on the railway pattern.

Looking through old carriers' lists is a fruitful but rather sad proceeding, for all too many of the places listed are no longer served by water let alone by boats. John Howell was once pleased to offer services 'regularly' from Birmingham to Measham and Ashby, Pickford & Co. to Newcastle, and Leek and Simpson, Hyde & New would be pleased to accommodate you on Tuesdays, Fridays and Saturdays to Derby, Cromford or Malton, all by water. George Swain went three times a week to such places as Melton Mowbray, Market Rasen, Pocklington, Horncastle, Cromford, and Pinxton. German Wheatcroft were pleased to state that boats left 'every Tuesday, Friday and Saturday from Derby to their Establishment at Buckland Hollow, whence goods are instantly forwarded'. John Whitehouse's fly boats went to Shrewsbury while Worthington & Co. operated services, in conjunction with other carriers, to St Helens, Lancaster, Kendal, Ashton, Oldham, Barnsley, Bradford, Bolton, Bury and Halifax. At Shardlow, George Moore ran the pub, acted as the farmer and was agent for boats. At Pinxton Messrs William Eperson of the Boat Tavern and Joseph Riley of the New Inn would be pleased to carry your goods, provided of course that it was out of hours.

Some carriers were rather vague, running occasionally or as the demand was felt, others were more particular. From Leeds to Selby the Aire & Calder Company boats were offered every evening 'Sundays excepted' (naturally) 'where they arrive in 12 hours and deliver their cargoes on board a steam packet which sails every morning and lands packages at Hull at the same p.m.' The Union Canal Company ran from Leeds to Liverpool daily, in four days, with '40 vessels engaged'. From Dewsbury, Joshua Kenyon left daily '12 at noon', from Huddersfield, Widow Welsh & Sons were pleased to operate to Manchester 'daily in 24 hours'. From Ripon to Hull, Martin Keddy & Co. would carry your goods twice a week 'in one bottom'. The bugbear of water carriage in later days at least, was that goods often had to be transhipped with damage and loss occurring.

A rather curious footnote to the above list is that 'The market boats which attend Hull market once a fortnight generally come with goods and passengers on that market day (Tue or Fri) which falls nearest the full or new moon, being the period of spring tides, the markets on such days are thence called "full boat days".' Obviously the intention here is that the full moon should have its well-known effect on ladies at the sales.

7 *Maintaining the Status Quo*

DOCKS & INLAND WATERWAYS EXECUTIVE   SOUTH EASTERN DIVISION

LOCK GATE PROGRAMME 1950

Lock site and No.	*Estimated cost Material*	*Labour*	*Stoppage*	*Total cost*	*Material in stock*	*When carried out and remarks*
Hawkesbury 1.	£89. 15. 4	£43. 10	£74	£207. 5. 4	Yes	July/August New Top Gate. New Timber Fits. Repair to lock side coping. Will mean two stoppages to complete. Work to be done on Sundays.
Hillmorton 4.	£55. 10. 0	£45	£65	£165. 10	Yes	July/August Fitting two new timber breasts to the existing bottom gates. Gates will be taken into the Yard for the work to be done. Refitted when ready as the traffic will not be delayed at these double locks.
Banbury 29.	£220. 15. 3	£88. 12	£75	£384. 7. 3	Yes	August New double bottom gates. New timber fits.
Heyford 35.	£209	£70	£69. 10	£348. 10. 0	Ordered Dec. 1949	August New single bottom gates. New timber fits.
Watford 3–4	These three locks require cast iron hollow quoins to replace old timber. Gates will stand over for another time.					
4–5	Work could be carried out as traffic permits, and delivery of new hollow quoins.					
6	It is estimated that the stoppage cost per set would be £80.					

'A stoppage of traffic (up the canal) is a thing most to be avoided.' This now rather trite statement was first written down in 1840, but stoppages are unavoidable and can be divided into two types. The first is the unprogrammed stoppage, usually brought on by the weather or by bank erosion: 'July, 24th 1769. Early on Saturday morning last, a little beyond Winsome Green [*sic*] in the Birmingham Canal Navigation the Earth fell suddenly in and Killed John Lester, one of the Workmen, occasioned, it is thought, by the heavy Rains on Friday Evening.'[33] A straightforward accident may occur, as recently when an English elm fell into the Welsh Canal near Irish Bridge just prior to the Easter Bank Holiday – a singularly inopportune time!

The other, an example of which is shown above, is the 'programmed stoppage'. Prior to nationalisation it was traditional to hold these over the Bank Holidays. A gang of men, up to a dozen in number, reinforced by boatmen who had perforce to lay idle, would work a 12-hour day, or, should conditions demand it, shifts through the period until the job was done, irrespective of the complexity of the works.

In the cases shown above it will be seen that the stoppages were held in mid-summer, giving the advantages of longer daylight hours and more or less guaranteed weather. With the traffic emphasis having changed to summer-season-only usage by pleasure craft such stoppages would be anathema to the governing body of British Waterways and programmed stoppages are now held between mid-October and mid-April. For both workmen and the quality of the work done such a change is unfortunate. During a stoppage at Long Buckby on the Grand Union Main Line (Whilton Flight) in the late sixties, conditions were so appalling that men found it necessary to shovel snow from the site before work could commence. Even the local cement company demurred at providing ready-mix, as their waterpipes and sand-hoppers were frozen. The fact that the job was finished on time speaks well for the stamina of the men concerned, although the economics of such work carried out in the period of shortest days and poorest conditions probably do not bear examining.

The works to be carried out on such a lock stoppage vary

from lock to lock and canal to canal, and one interesting variation concerns the presence, or otherwise, of the foreman. In general, British Waterways Board foremen are ex-craftsmen 'made-up', and if they were interested in stoppages they would be present, if not, then probably they would be occupied elsewhere; in the latter case the responsibility for the work tends to fall on the (wide) shoulders of the carpenter. Until the early 1950s it was usual for lock-gates to be made at the section yard, an involved job which could be guaranteed to keep the carpenters warm throughout the winter. A form of lock-gate, the existence of which is often overlooked, is the so-called safety-gate, which is fitted at the end of a valley or where the canal is high above the countryside on an embankment. A typical waterway utilising these is the Worcester & Birmingham Canal, where along the summit there are eleven sets. Basically they are like a pair of lock-gates, often situated underneath a bridgehole, and the theory is that if there were to be a sudden drop in water they would 'come to'. Unfortunately, due to the vandalistic propensities of some individuals, it is necessary now to keep them chained back, although there have been occasions when these chains have been sawn off with singularly ill-effects upon craft which did not anticipate and could not see a pair of strong wooden gates just on the waterline. All these gates were closed during the last war, when bombing was in progress and, in fact, the aqueduct at Bournville had a direct hit from a bomb. Had the gates not been closed a large area of Bournville and Stirchley would have been washed into the rivers Bourne and Ray. In 1972 it was necessary to de-water the one-and-a-half-mile-long pound between Kings Norton and Cadburys, Bournville. For this purpose the safety-gates were utilised to retain the water, but, far from being instantly usable, work commenced on clearing them on 28 January and continued thereafter until the actual de-watering on 14 February. About 100 tons of rubbish was removed, necessitating the employment of a dredger-driver, discharger-driver, tug captain and various labourers, together with the relevant craft. To compound the loss of time both the dredger- and discharger-drivers, men in their mid-60s, were taken sick; there was no one else on the section who could drive either machine so a man had to be borrowed from another section.

In many areas old methods and equipment have lingered far too long on waterways. The hand method of scooping water from the lock chambers continued until relatively recent days, and in fact is the only way to put the finishing touches on such work today, but where there was a long pound a manual pump was utilised around the turn of the century. Made of timber and iron it would be lowered into the lock by stop planks, and a stage fixed for the men to stand on. After priming the pump six men would start swinging the handle and pumping away. After half an hour it was, of course, necessary to relieve them with another six men and so on, until the lock was empty, which might take anything from two to four hours. It is not surprising that songs and ditties were sung to keep up the rhythm. These were often sea shanties, and if the words were not generally known one man would set the pace by calling out the words.

A typical steam pump which came into use during the First World War necessitated iron suction pipes, the fitting of these being described as a 'heavy mauling job'. Fitted to a boat, this craft would be floated up to the bottom end of a lock and the overall time required to pump out the chamber was reduced to one hour, providing that the stop planks fitted tightly and were kept well racked with fine ashes.

The next improvement, at least on the Worcester & Birmingham Canal, was when the superintendent procured an engine out of an old car from the local schoolmaster. This was a very early vintage $2\frac{1}{2}$ h.p. De Dion, mounted on a chassis with iron wheels, and an improvement albeit still with iron suction pipes. Later, yet another pump utilised an Aster engine, also from an old motorcar. The first pump with flexible suction pipes, following the local trend on this waterway, was powered by a BSA engine removed from a car!

Mechanisation on many waterways also reached the sawing division around the turn of the century. Prior to this all such heavy timber cutting was put out to contract, being sawn by roving gangs who used an enormous crosscut saw, one man working from above and the other standing in the pit below the timber being cut, the man above making the lead stroke and ensuring the saw was kept square. One essential requirement for these men was plenty of cider and beer.

The bricks used on a waterway normally came from local sources, but if the canal fell into railway hands or the local brickworks closed down some variety can now be found amongst the coping stones. Locally, a half-mile stretch shows bricks dated between 1870 and 1900 from Joseph Hamlit, West Bromwich, the Earl of Dudley's Coneygre brickworks at Dudley Port and from William Morris of Oldbury.

Typical of the maintenance work carried out by bricklayers is the widening of weirs or, in recent times, when a stop lock has been done away with, raising or lowering them throughout the length of a waterway. It will be understood that on a summit or any intermediate level should one weir be lower, even by so little as half an inch, than the others, it will always be running and wasting water. In areas where there is extensive mining, subsidence often necessitates the raising of weirs, on some occasions as much as 3 ft over a number of years. The Wyrley & Essington Canal as built was a contour canal, keeping to the level of the surrounding ground, but this ground has subsided over the years and the canal banks, having to remain constant, are rebuilt high above the neighbourhood. The bricklayers' handiwork is very apparent here.

Like most canal tradesmen, bricklayers have not necessarily served an orthodox apprenticeship, working instead with different older men until the trade had been learnt. For the first two years of his working life one (now retired) bricklayer acted as the general tea boy for the gang he worked with. His first job in the morning was to sweep out the boat and have the 'kittle a-billing' and the tea cups, not mugs, together with the tea pot, sugar and milk, laid out ready for the men when they arrived. For the extra time required he was not paid by the company, nor by the men, but he was rewarded – it was required that he buy all his own tools, but instead of having to do this, as a recompense for his services, the men gradually built up a collection of tools for him. The cabin of their boat, which it must be stressed they had regularly and exclusively, was painted blue and white with carpet laid on the floor and palliasses on the benches. No man was permitted to enter the cabin with muddy boots – relics of this practice are still to be seen among older hands today – the men carefully wiping their feet before getting on board. Each man had his own cup and saucer and paid 6*d.* (2½p) a week into the common tea fund, the lad being responsible for all purchases. However, sugar was 'freemans', as this was a traffic operated on the waterway, while milk was obtained from local farmhouses in exchange for coal, acquired by some means. While the men had their tea the lad would get the tools from the front cabin and lay out those he thought each man would require. When they had finished their tea he would, again, sweep out the cabin, wash up the cups and saucers, hang the cups on their hooks and generally make himself useful. Each man brought his own 'bait' in his 'snap tin' for lunch, some would have a hunk of bread and cheese, or cold bacon, others would need their food cooking for which the lad was responsible – on the understanding that there was never any need for him to bring his own lunch. One or two of the men, he recalls, would go up the bank and bring back a rabbit, which he had to skin and paunch ready for the men to take home. As the years went on he was promoted and the new lad took over his duties. Rather curiously the last job he had before retirement was to rebuild a pump house, the original of which he had helped to build 50-odd years previously.

Before commencing the serious, heavy, work of carpentering on a waterway, the younger apprentices would make, in a matter of months, such items as 'horse and donkey supporters' to be handed over to the saddler to fix in the harnesses; water scoops; wheelbarrows in two sizes, small for navvies and larger ones for dredgers; ladders, short ones of 25 rungs or longer ones of anything up to 36 rungs; house and cottage doors; window frames and little 'odds and bobs' including painting and figuring the gauging sticks for the toll clerks. Among the odds and bobs one lad '. . . used to have to take the men's time sheets and works repair report every fortnight, usually on a Tuesday afternoon to Stoke Prior (walk down the canal-side from Tardebigge), carrying a locked leather dispatch case containing the time sheets, etc., and deliver to the Engineer, a Mr. W. Griffiths, who lived on the wharf at that time'.

38 Opening out Armitage Tunnel, Trent and Mersey Canal, 1971

Blacksmithing was another trade essential to good maintenance on waterways. Iron or steel being bought in, in the rough, rod, flat bar, square bar or tube, as needed. A day's work might be the making of paddle rods, these having to be heated at one end until they were red hot, flattened and a slot cut into them to allow for the insertion of the cotter pin which retains the rod within the rack mechanism. When they had cooled, the other end would be tackled, a flat bar, which was bolted to the paddle, being forged (latterly welded) on to the rod. We are told that it was far easier for a skilled blacksmith to knock a 1 in. circular hole into a red hot piece of steel $\frac{1}{2}$ in. thick than it is to drill it nowadays. New cranks would be needed for the steam tugs, castings of racks, pinions and safety catches had to be cleaned up and made to fit. Bolts would be manufactured as required. A bugbear to modern maintenance workers is the fact that all too often when a replacement nut or bolt is required the existing one, far from being a Whitworth or BSF standard thread, is one produced by a set of taps and dies which some long dead blacksmith had made for himself. As a side-line the matching of boat hooks to shafts would be carried out. A weird modern tendency, failing a stock of old-fashioned rivets, is to buy and cut down coach bolts, leaving the head and such shank as is required for the job, the end merely being peined over. This is hardly an economic or craftsman's procedure, the blacksmiths of yesteryear must turn in their graves. In fact as far as the blacksmith was concerned nothing should ever be 'put out' that he could possibly, one way or another, codge up in the workshops. There are still a few blacksmiths on waterways, carrying out this honourable trade and one we know, far from being the burly strong man of fiction, sweating madly over his forge, is, instead, a small, loquacious, happy man who does not rely so much on the strength of his back as on the skill of his fingertips.

A latterday requisite of at least one canal maintenance man was a knowledge of electricity, in so far as he was required to maintain the private telephone system operated by most independent waterways. It is said that when the National Telephone Company came into being they rode around to most waterways suggesting the installation of telephones here and

39 Breach at Bollington, Bridgewater Canal, August 1971

there, mainly to speed up communications between the lock-keepers and the administration; but upon requesting and receiving an estimate a cold shudder ran down the spines of the powers-that-were and instead, and until quite recently, they ran and maintained a private network. There were even hazards in this:

> 'The telephone wires used to be fixed under the tunnels at one time, I remember I was helping to keep a boat steady in a tunnel, while two men were repairing an insulator fastening, when one of these men suddenly fell on to the cabin-top, where they were standing to reach the work to be done, we did not know what had happened to him at the time but thought he had fainted or something. However, we went out of the tunnel to try and revive him, which we eventually did, and it appeared there had been a thunder storm while we were in the tunnel and he had had an electric shock.'

Nowadays there are a number of skilled tradesmen whose job is often underrated. Among these are the electricians whose jobs are infinitely varied, re-wiring the loom of a dredger miles from anywhere, which for reasons unknown had decided to burn itself out, being one minor example. Another specialist trade is that of the fitter who may one day be rebuilding a Petter, Armstrong Siddeley or Lister diesel engine in a working craft and the next day take an emergency call-out to a pleasure craft in trouble with hydraulic failure. Later he may be tinkering with the bucket on a dragline and after that installing a new stove on a motorboat. Incredibly, most are good-natured.

Most of the lorry drivers employed by the British Waterways Board have many encounters and meet many hazards. Looking at a canal with the weird tracks leading to old wharves, weary old bridges and barbed wire, it is little wonder that being mired in is 'just one of those things'. One wonders, though, how the driver of one Waterways' lorry felt when he collided on a narrow bridge with another Waterways' vehicle – that accident report to headquarters must have made interesting reading! It should be borne in mind, however, that they also manage to drive upwards of 200 miles a day, come rain, come ice, or sun.

Some other drivers employed with the BWB have, more or less, replaced boats' captains; for in all too many cases all traffic to or from warehouses is handled by road, and the trucks operate on a scheduled run, day in day out.

Piling is normally a two-man job under the guidance and assistance of a ganger. Materials have varied from plain wooden piles driven in by hand to pneumatically driven interlocking convoluted steel models. A transitional phase incorporated concrete piles driven in by a heavy hammer operated from a boat by means of a winch. The basic uses of piles were always to prevent undermining of banks in cuttings and bank-strengthening on embankments. Here the marked contradiction between the commercially viable waterways of the north of England and the moribund canals of the south and midlands becomes most apparent. As one potters along the Aire & Calder, neat concrete coping and a firm bank are visible, while the towpath of the Staffs & Worcs canal looks like Old Man's Teeth, with gaps and broken pieces abounding. One ex-ganger, talking of piling, put his finger on the hub of the matter: 'You see, if you got piles, the banks don't fall in, and you don't need so much dredging.'

Canal painters have always been in existence, although often a man would double-up on two jobs, the carpenters, for instance, being quite capable of looking after the painting of their own work. In keeping with modern trends such painters as waterways have appear to do little painting, most of it being put out to contract; instead the painters paste up more notices to become defaced. A fairly common grouse today from the older men is that: 'When I were young, you dain't need to be told about a job, you done it, there were no jobs we carsn't do'; whereas the newcomers to waterways prefer to know exactly what they are expected to do and when they will be doing it. Demarcation, however, does not in any vast extent rear its head, but on the other hand the influence of the unions on waterways has been beneficial to the working man, and in 1971 a grading system was introduced which laid down the jobs a man could do within his rate of pay.

A labourer might be expected, over a period of a few weeks, to undertake hedge-laying or brushing, hole-digging and leak-stopping, boat-steering (eel-catching), off-loading bricks from a lorry on to the bank and, a few days later, loading them back on to another lorry, handling a pneumatic drill, oiling locks, pumping out sunken pleasure craft, general stoppage work, stringing up a new wire fence to replace a hedge maliciously burnt down, ditch-cleaning, dislodging any unwanted articles from culverts, burying dead, decomposing and/or liquefying carcasses of animals (at 10p per head), clearing overhanging growth or one of a dozen other jobs. Some jobs, lorry-driving or boat-steering for instance, require a modicum of intelligence, others not. The differential between the one and the other was 10p per week in 1972.

Of the men employed a fair percentage will be at or over retirement age, content to scratch away in an obscure corner of the waterway, working conscientiously, if slowly, at some mind-numbing task that could be done in a twentieth of the time with a machine; others, by virtue of their friendship with the foreman or other official, will be found yet another obscure corner of the waterway where they will contemplate a space 6 in. in front of their noses while a few will have found themselves a tranquil niche in the hierarchy of the waterways, often having opted out of the pecking order in exchange for a job which can be made interesting. A number of men, age is immaterial, are enthusiastic about the job, but the older they are the more competently it is carried out. For example, a willingness to spend 12 hours of a night checking the water levels into a reservoir in times of rain, to go out in a van at two o'clock in the morning with a motor pump in order to help a pleasure craft owner whose boat is in danger of foundering after hitting an underwater obstruction, or to abandon the television at half past nine of a summer's evening to ascertain and check the damage done to the parapet of a bridge by moronic personages, is found amongst these men. They are the type of men who, when out cruising on their own boats – and some canal workers are, incredibly, able to enjoy a busman's holiday – will endeavour to rescue a pig which has fallen down the bank into the canal, or will spend their spare time writing a report on local birds for an ornithological society or upon canal life for a church magazine. This enthusiasm, rather strangely, is more apparent in the lower strata of maintenance men than above; too often, due to the system of promotion (dead men's shoes), the man above has become blasé on the whole subject.

A class of men whose enthusiasm, or lack of it, can make an enormous difference to the work carried out, is that employed on dredging. Here there is no clearly defined approach to the work, some gangs working on normal rates of pay and plodding away, others on the same rate rely upon the 'tat' (brass, copper, wire) that is to be found in the bottom of the canal to boost their pay, others may find a bonus scheme applicable, whereby over and above a certain tonnage an extra sum is put in the pay packet every week. Dredging, by governing the depth of water, represents the tonnage capacity of a waterway and since early times various technical advances have been made on the machines used to keep pace with the ever-increasing accretion of materials hidden beneath the waters. Foremost of them is mud – mud brought down to the waterway from the sides of cuttings, from faulty drains and freshets or rivers. It is unfortunate that a canal is dependent upon these natural sources of supply to maintain its water-level and that the one must go hand in glove with the other. What are extraneous are the polythene bags, mattresses, bed frames, lamp standards, telephone cables, diesel and waste-oil drums, tyres, prams, suites of furniture, carpets and the like which arc so carefully deposited in the canal, not only by individuals but quite often by concerns who should know better. Within a twelvemonth it has been recorded that Post Office employees disposed of some unwanted cable, a motorway contractor dumped ten cubic yards of rubble, a chemical firm mistakenly ran the wrong pump into an outlet, killing all the fish thereabouts, didicoys pumped upwards of 50 gallons of waste-oil over the bank and contractors for a local council disposed of broken paving stones over a canal bridge; this was all seen over a 20-mile stretch. Whether the strengthening of anti-pollution Acts in 1972 has had any effect is doubtful and until these people and firms can be caught and used as examples it is unlikely that any Act of Parliament, however strong, can materially alter ingrained habits. A flat statement is that a boat drawing 3 ft 6 in. of water can pass through a bridge on the outskirts of Birmingham one day and be stemmed up the next. A week later a dredger will clear the obstruction, but within 24 hours the position will be back where it was.

No definite date can be fixed for the introduction of any one type of machine used for dredging, but the grand-daddy of them all, and he is still alive today, is the bag and spoon – simply a leather or canvas bag having a circular ring of iron around its mouth, the ring being fastened to a long pole. One man in a barge guides the pole, the other drags the bag along the bottom of the channel by means of a chain at the bottom end of the pole, which is wound around a winch at the far end of the barge. The mud is scooped up in the bag, the bag raised, tipped out into the barge or hopper, dropped back down again and so on *ad infinitum*. An aqua mortrice is an improvement on this method, utilising an iron scoop in lieu of the bag, this iron pocket being wound up by a small motor. It was estimated in 1896 that 65 cubic yards of gravel could be raised within a working day of 12 hours with this machine and it was probably an ancestor of this that was in use on the River Severn in 1842:

> This town [Upton-upon-Severn] owes its chief importance to the river, which is here navigable for vessels of 110 tons, and is being much improved by the Worcester & Birmingham Canal Co., during the present summer, who have expended nearly £2,000 in procuring a dredging machine, worked by a steam-engine of 8 horse-power, which brings up from the bed of the river about 150 tons of gravel per day; and it is hoped, by thus moving the shoals and banks, to render the river completely navigable at all seasons. If the silt thus obtained from the shallow places be judiciously deposited on the low banks, where the river is widest, there is no doubt much permanent good may be done to the navigation by narrowing and deepening the channel, in the way the Clyde and other rivers have been improved.[34]

For use in the wider, deeper canals there came next a bucket-and-ladder dredger. An eyewitness report of the operation of one of these on a narrow waterway in the early 1900s gave the method of operation very clearly:

The dredger was iron hulled, fitted with elevator chain of buckets, one mobile steam engine, this engine was lifted out of the dredger when the dredger was not in use, or if the engine was required for more urgent jobs such as power for a centrifugal pump, etc., the engine would be lowered on to its chassis to be towed away to wherever it was required. There was a special bed in the dredger that this engine was lowered on to, so that the steam pipe and water intake pipe could be connected in a short time. When steam was up to required strength the elevator chain of buckets would be lowered on to the bed of the canal or the silt to be moved, the driving belt fixed from the flywheel on the engine to the pulley shafting driving elevator, the buckets loaded with silt or mud would rise up from the canal bed with the elevator in motion, the buckets would deposit the silt or mud on to a point. [The dredger would load a boat either side of her, whichever side was suitable to load.] As they went over to return to the canal bed again, this silt would slide down the point and into the iron container that was in the boat alongside the dredger. There were four of these iron container tanks in these boats, they would hold five to six tons of silt. Donkeys were needed to tow the loaded boats to the tip and bring empty boats back to the dredger to be loaded again. The procedure at the tip – the steam crane worked on iron rails similar to railway lines, the rails being laid down on timber planking, and trestles where required. The steam crane worked along these rails under her own power, when a loaded boat came alongside the tip the crane would move up to the side of the boat and the lifting chain would be fastened to the lugs on the sides of the container tanks and lifted out of the boat and taken by the crane to the end of the tip, a stop lever being knocked out of its closed position, causing the tank to turn, depositing its load into the tip, the tank being righted again was taken back by the crane to the boat. This procedure went on day after day from dredger to tip. Many tons of silt and mud would be taken out of the canal every day, depending on the distance from dredger to the tip. A forty-foot long boat was used to stable the donkeys at night, the punt was required to take chains ashore to fasten the dredger to mooring stakes, otherwise she would crawl her way along when working. This was called the chain punt and this punt would also be used to fetch coal from a convenient place where coal had been placed for the dredger. There were four men required to work this dredging plant. The last time this dredging plant was used was in 1912.

The above refers to the dredging of a new winding hole, to enable the narrow carrying craft to be turned more easily. Unfortunately due to a lack of communication, or inattention on the part of the local lock-keeper, on a night of heavy rains excessive water was run off, causing the dredger to tilt over. As she rolled so the water ran over the low sides and she sank, though square on the bottom. In the process of trying to lift her one of the chains broke and she fell on her side in the middle of the fairway. To get her out a dam was driven across the canal and that immediate length of waterway was drained; although she was eventually righted and taken to the workshops she was never used again. When, during the First World War, she was broken up, she was still in good condition, many of the plates still having the red lead paint on the bottom where the ballast had been laid. Although always regarded as being before her time, she had, rather incredibly, to be bow-hauled by men wherever she went. The mobile engine, however, ended up functioning in a saw mill, supplying power to the circular saw used in lock-gate making, and was not scrapped until 1924.

The principle of a dipper bucket dredger, a near relative of the bag and spoon, served both for drag lining and stray dredging. Essentially it is a large cylindrical steel bucket with a projecting lip fastened to a crane, jib or boom, which is lowered into the canal and, by means of chains, drags itself into the mud, the jib raising it above water level. The bucket itself can either be bottom-opening or capable of being tilted to deposit the material. A grab-bucket dredger is the type most commonly in use now. The principle is that the grab is suspended from the jib of a crane and is lowered with its jaws open to the bottom of the channel, penetrating the surface by impact and its own weight. The jaws are then closed trapping a mouthful of whatever the grab has found; the whole is raised and swung until it is in alignment with the hopper or waste boat, the jaws opened and the muck deposited therein. Different types of jaws can be provided, ranging from smooth lips to heavy teeth. There is nothing more frustrating than to operate such a machine, lift, open and see one brick drop out! Alternatively, if a brick is caught in the jaws, the silt will wash out during the lifting and swinging motion, or if the bottom be very hard it is not unknown merely to lift a grabful of water – or fish! Early models were worked by a steam engine with a boiler but problems have always been met with the rolling motion resulting from the movement of the crane jib. The vibration and rolling necessitated the monthly docking of early wooden craft and the inherent instability has caused a number of such dredgers to roll over.

Around the turn of the century it was common practice on narrow waterways to use a 7-ft beam dredger, which was either held by wire ropes or chains to iron bars driven into the banks, or supported on each side by sealed pontoons, wherever the waterway width permitted this. By virtue of their design they were not normally self-propelled, but were punted around with a small steam tug. A large crew was needed, two boiler tenters, one crane driver, two men on the tug and various others operating the boats with the waste and the steam crane at the tip; their capacity was claimed to be 400 tons a day. Today's dredgers of this type are self-propelled, fitted with a diesel engine which also operates a hydraulic grab. Theoretically only requiring one man, two are found to be desirable if the driver is not to run backwards and forwards from the grab controls to the steering controls time and time again, although they can pull themselves along the bottom of channel by the action of their grab. The latest design is controllable from the cab, but at the insistence of the union an extra man is often carried for safety reasons. The problem of stability has been overcome in a most ingenious manner, which can be effectively described as four legs with large, flat feet positioned one in each corner. By means of hydraulic arms these are lowered to the bed of the canal under considerable pressure and are self-adjusting until the dredger is levelled. Accidents can still happen very easily; without the feet in position they are hopelessly unstable and have rolled over when the operator has omitted to lower them. If moored in a lock pound special care is paid to the level of the water, as they are incapable of floating themselves off the bottom. Large warning notices are distributed liberally over the structure and the actuating mechanisms have the legs painted a lurid colour, following an unfortunate accident when the son of a dredger-driver put his head between the hydraulic ram and the cross arm as the leg was being raised. The father subsequently left the employ of waterways.

A final type of dredger, which is worthy of mention here and is in use on some river navigations, is the suction or centrifugal model, of which a number of varieties are available. The silt is sucked into the innards of the machine, mud being thrown out one side and water the other. They are, however, incapable of digesting the detritus of our civilisation and for that reason are not likely to be extensively used on the smaller inland navigations.

The deposit of waste mud has, over the years, fallen into three distinct patterns. In the early days most farmers welcomed the slurry and rushes ('segs') pulled from the bottom of the waterway as a good fertiliser. The material was shovelled out of the hoppers into barrows and, using long planks could, by a judicious agreement 'twixt ganger and farmer, be taken quite a long distance to the required field. As artificial fertilisers became easily available, the material lifted contained increasing amounts of 'foreigners' and pedigree herds of cattle roamed the land, farmers no longer wished for such treatment, and it became necessary for the various waterway concerns to find among their own bits and pieces of land, or to purchase, a declivity wherein the muck could be dumped. In suitable positions a quarter of a million tons of slurry can be disposed of. Needless to say by its liquid nature this material can often run amok, but it is the proud boast of the crane operators responsible for the discharge of waste that thcy can turn the course of the tip with a matchstick or a twig. Certainly the odd railway sleeper applied at the opportune moment and in the right place can swing the course of the mud 100 yards to one side, thus giving earlier deposits a chance to drain and settle. An alternative but restricted method of disposal is to dump the material on banks to raise them, or to backfill new lengths of piling. The final approach to the problem, more practicable on river navigations, involved the use of bottom-opening hoppers which could be towed to an estuary or spit of land to be reclaimed, where the contents were dumped.

A typical modern operation on a midlands canal will necessitate the use of a dredger, two tugs, a discharging crane, and six or eight waste boats and any number of men from a minimum of four to a maximum of eight, exclusive of supervisory staff. In theory two of these men should be plant operators, one existing full time on the dredger at the dredging site, the other operating the crane at the tip. A further two men should be tug drivers and an additional two men are normally required, as a minimum, to steer the waste or mud boats. Still, in theory, we should have two craft being loaded at the dredger, two being discharged at the tip and two trains of a tug plus two boats operating up and down the stretch of waterway. Sometimes, obviously, when the spoil can be dumped locally, it is easier to omit the tugs and shaft the boats to their destination, however it will be understood that on a narrow waterway tugs cannot be turned at any convenient point, but must often have to run a mile or so at either or both ends for this purpose. Again, the tugs are rarely matching pairs and one finds combinations like a converted icebreaker of 1900 vintage running the upward leg and pulling two boats, while the downward leg is being pushed by a modern (1970) pusher tug. The dredger may be working in an awkward site or in poor conditions where it is incapable of loading the boats at the same speed as the crane at the top can discharge them. A high wind can make the marshalling of the boats difficult or the proximity of a pleasure-craft base may mean that they have to be moved about to a great extent, as the standard penalty for damage to a pleasure craft is three days suspension without pay for the crew concerned, irrespective of cause. Not the most enviable of jobs.

8 *Outside Influences*

> That the inland waterways are to have a board of their own is good news, though it comes long after it was due. The independence now being given to the canals must be genuine, however. That implies that they must be admitted to have a strong claim to a share in the money that the Exchequer is going to devote over the next five years towards trying to straighten out the nation's chaotic transport system. So far the waterways have been the poor relation of the transport system. They deserve enough assistance to ensure that the results of administrative as well as physical neglect are overcome, for they still have a very important part to play in relieving traffic on the roads.[35]

Canals as a form of transport had a very brief hey-day, probably not more than 50 years. Thereafter their very existence was, and still is, subject to the whims and wants of external demands. Briefly, the railways bought some, closed many, but those they retained were worked hard. Of the independents some – for example the Wey & Arun – just died, others fought on as will be shown later. When the First World War came the canals and their men did their bit, but some never really recovered from the lack of maintenance necessitated by shortage of manpower. Those left within the railway scope were run on with less and less enthusiasm, both the GWR before the Second World War and the LMS during that war quietly getting rid of the bits and pieces that 'didn't pay'. Then with nationalisation came the last chance for commercial waterways to become a viable proposition. Instead it is on record that one well-known member of the D & IWE when told that canals were a part of his problem exclaimed, 'Oh Lord! do we have to have those as well?' This attitude, sadly, had not, according to the *Birmingham Post*, changed in 1958:

> Third time pays for all but Mr. Job Clarke, a canal waterman who took the last narrow boat out of Stourbridge Basin yesterday. He has done this kind of thing before; he took the last boat along the Wolverhampton Canal, and the last one from the old Brierley Hill Basin. Each time he managed to find himself a berth on another section of the Midland network of waterways. But this time it looks as though he has finished with the canals for good, as he is being found a job as a railway porter. This after a lifetime on the water. Mr. Clarke, who is 61, comes from a family which has worked on the canals for over 150 years, and he was born in a boat alongside a canal wharf in Birmingham. The section of the Stourbridge Canal which runs to the Richard Thomas & Baldwin Works at Swindon, was, until quite recently, carrying about 150 tons a week. Now British Waterways has found it 'uneconomic' and the job will be done by lorries. Three other boatmen will also be found other work by British Railways.

In 1963 docks and waterways were made independent of railway (British Transport Commission) control – a move bitterly regretted now – and the emphasis was, and is, in so far as southern and midland canals are concerned, solely on pleasure craft, sufficiently so for this phenomenon to be discussed in chapter 9.

Fundamental causes of the decline of the waterways are not

difficult to discover. In some ways most never recovered financially from the loss of confidence by shareholders when presented with what was, on the face of it, a far more efficient and economic form of transport, the 'Rail Road'. This loss of confidence led in due course to a parsimonious attitude to improvements, or indeed to any maintenance work.

Canals, as cannot be stressed too often, are dependent upon heavy goods for their livelihood. Such ephemeral items as market boats, packet boats and lightweight goods boats, whilst providing a welcome dollop of sauce could not provide the meat, or revenue, for shareholders. Where one or two projected canals were concerned a cold douche of realism was necessary.

In 1825 the opposition to the proposed Romford Canal asked:

> Is it to carry agricultural produce from a large extent of country, *where there is no market*? No: for Grain often fetches as good a price at Romford, as at Mark-lane. Are there any manufactories of a *staple commodity*, at or near Romford, saleable in London, or elsewhere, and admitting of a profitable export trade, any spinning and weaving of cotton and wool as at Manchester, and in certain towns of Yorkshire? No. Are there any collieries, or mines, as in some other counties? Neither one, nor the other.
>
> It is said in the Prospectus of this Canal, that manure and coals would be more readily and cheaply conveyed by water: but no man can *reasonably* expect that the freight of these things would ever pay common interest to Subscribers for the sum required in this undertaking.
>
> If an example be required, take Croydon, which is at about an equal distance from London, *also in an agricultural district*, where coals, lime and manure are required: but this had the advantage of transit of goods to and from the extensive bleaching grounds in the neighbourhood. The £100 shares sell for £4. 10*s*. They have paid no interest for many years. . . .
>
> The same at Basingstoke and at Ashby de la Zouch. The Thames and Medway Canal has *four* loans; but the *original Share-holders* receive no annual dividends.

Let any one ask at *Salisbury* about the Canal, and he will be told, that all the Instalments upon the £100 shares were paid up, and the money expended; that a call of £50 more was made upon each Share-holder, that some of them submitted to the first loss, whilst others advanced this additional sum, *which was applied to liquidate arrears*, and the work was abandoned, about six or seven years ago; yet there must have been plausible arguments employed by the Proposers of all these undertakings, and prospects of great advantage held out, to induce people to invest money in them.

Independently of such considerations as arise from the mere hazard to Proprietors of land, or probably loss to Subscribers, there are a great many persons, unconnected with Land-holders in the marshes, who object to a Canal, because it would *inevitably* lead to the augmentation of Rates, increase the number of parish settlements, by those employed in digging the Canal, and working barges upon it, (Excavators and Navigators) and . . . in fact, the only persons likely to derive any advantage, would be those, who were connected with the executive part of the enterprise, *as long as the money might last: but the Injury would be permanent.*[64]

That which was relevant to Essex was just as relevant to the rest of the country. Disregarding the apparent deliberate policy to run down the waterways, agricultural traffic could not have maintained the canals even in 1800 and much less so now.

So-called 'smokeless zones' have meant a changeover, for even if solid fuel is still used it is that from the Welsh pits, anthracite, and not coal from the canal-side pits in the Midlands. Possibly even this might have been overcome had coke still been processed from old-fashioned gas works; instead, as natural gas from the North Sea becomes more widely used, so sources of coke dwindle. Also a large number of private dwellings and factories have changed from solid fuel to oil. This could to a certain extent have proved advantageous to canals for, certainly in the larger sizes, no road-tanker can ever compete with a waterborne tank-boat; but pipe-lines, the chosen alternative, while paying a wayleave for permission to pass along the towpaths, do not help to keep the channel clear.

When the canals were first built warehouses and factories sprang up all along their banks, now these factories are often scheduled for demolition and their replacements are being built elsewhere, often on good food-producing ground – the demolition sites remaining an eye-sore until someone can offer to buy them from the developers.

There are other petty irritants, petty only in their magnitude, which have put the finishing touch of death to many canals, at least commercially. One example is the Bridgwater & Taunton Canal which had swing-bridges, now they do not swing as they were fixed in the closed position during the war, ostensibly as an anti-invasion obstacle. Public apathy has ensured that they remain fixed.

The last and final straw is the growth of motorways. Canals have been closed, the top end of the Lancaster Canal is a good example, because it was more convenient for the motorway to pass over a culvert than to build a decent bridge. Other canals have, by the artificial restriction placed upon them by the supporting piers of the motorway, had their dimensions fixed for life at the old narrow gauge. The Birmingham Canal at Oldbury whose line was altered to accommodate the motorway, is a good example. It would take some vastly expensive operation to widen it to a 100-ton capacity, let alone 350 tons, the minimum which our partners in the European Common Market find to be desirable, and so commercially it is finished.

Another development, much praised, has been that at Farmers Bridge, where in exchange for the loss of a few wharves, lengths of canal and general loading and unloading facilities, a pretty public house has been built for the use of pleasure craft owners.

Now it will be understood that the British Waterways Board's maintenance engineers are overworked and suffering, in the lower echelons, from gross undermanning, so it would be unreasonable to expect the narrow canals to receive anything other than vital works. Such trivia as weed-clearing and rubbish-removal takes place, if at all, as and when possible. The City of Worcester had a centenary celebration in 1971, and panics

40 'A stoppage of traffic (up the canal) is a thing most to be avoided.' Braunston, Whitsun 1922

41 Ellesmere Port docks and warehouses, burnt out 1 April 1970

42 All that is left of the Thames & Severn Canal lock at Chalford

43 (*below*) Compartment boats with diesel tug, Aire & Calder Navigation, *c.* 1970

ensued when it was realised that the canal at Lowesmoor, through Blockhouse and down to Diglis was in a bad state, so men were drafted in to clear it up. Not, you will understand, to dredge out the city's refuse that had been carefully deposited in the channel, or even to paint up the locks, but to clear the weeds from the edges and cut the grass and bushes along the towpath. They even had to clear away the cider and methylated-spirit bottles left by an itinerant gang of Irishmen who had frequented the towpath thereabouts. Did Worcester look any better for having its canal tarted up? – not it!

The many causes of the moribund state of most waterways interlock, but certain threads can be extracted from the tangled skein. It must have been a very uneasy time for the canal shareholder when railways, as opposed to tramways, were first mooted. Possibly they needed a schizophrenic mind to judge which way to jump. The average canal shareholder possessing a reasonable number of shares and a modicum of intelligence was already aware of the advantages of tramways. Even if, and it was rare, his own canal had no tramways, he would have seen them in use, feeding from a mine or to an iron mill. He would have noted that although it was not, in terms of bulk carriage, as efficient a form of transport as his waterway, nevertheless the early problems had been eliminated. Benjamin Outram or Josiah Jessop have been variously credited with taking the edge away from the earliest L-shaped rails and substituting instead a flange on the outside, or inside, of the cauldron's wheel. However, tramways were still not entirely cut and dried in their design, although the practical gradient for use by horse-drawn cauldrons, or wagons, was known and more or less fixed.

> On the edge-railway, the rims of the carriage-wheels are made broad, with a flanch on the side of the wheels to prevent the carriage from leaving the rails, which renders the carriage or *wagon* considerably heavier than the tram. On the other hand, the wheels of the tram are narrow in the rim, coming nearly to an edge; and there is a flanch on the inner side of the plate upon which they moved, to prevent the carriages or trams from running off the road.[36]

'Again, I completed a tramroad from the bottom of an inclined plane near Dowlais Works to the navigation house on the Glamorganshire Canal, a distance of upwards of nine miles, the fall on some parts of the road being upwards of one inch per yard, and on the remaining portion varying from two to six inches per chain. Upon this road, like the preceding, a single horse continued for some time to haul ten tons, and bring the empty trams back, travelling regularly a distance of nineteen miles per day. The horse here referred to, like those mentioned above, was greatly inferior to the horses generally used in the north.'[36]

The wagons themselves had also achieved, with local variants a compromise set of dimensions and capacity.

'I will here observe, that I have tried every species and shape of trams; and among such as are made of wood, designed for use both under ground and above, I have found none equal to those which are constructed with sides one foot upright, and nine inches upon that, to bevel a little outwards. The sides consist of four elm planks, two on each side, the lower planks two inches thick, the upper ones an inch and a half, with straps of iron under the bottom, extending to the top edge of the sides, which are screwed together with inch elm boards.'[36]

The animals used on such wagonways were overworked and it must be borne in mind that a canal horse could pull up to 50 tons without excess strain.

As the horses belong to the waggon-men, whose occupation is not much more profitable than that of a common pit-man, it may be supposed they are not excellent cattle, nor in high order; yet such is the advantage of draft on these Rail-roads, that they usually convey a Newcastle chaldron (about 50 cwt.) to the river side. If the coal-waggon be supposed to weight 10 cwt. no less than three ton is thus easily transported by one horse.[37]

The loading and discharging of such wagons was well

organised, nowhere more so than in the area of the northern coal mines. 'When the waggon arrived at the Staiths, he lets down his Bottom-board, which is the bottom of the waggon, having Hinges on one side, and a hasp on the other; and the coals run down an opening in the waggon-way under the waggon, which has a box projecting off the dike of the Staith upon the water, under which the keels are placed which receive the coals . . .'[37]

Obviously, even accidents had reached a point where everyone was rather blasé over the matter.

> Waggon-men, in going down very steep Runs, always take their horses from before, and fasten them behind their waggons; as they would inevitably be killed was the convoy to break (which frequently happens), or any other accident occasion their waggons to run amain; nor is this fatal consequence only attendant on the horses, but the drivers often receive broken bones, bruises, and frequently the most excruciating deaths. Indeed, in some place, a most humane custom is established, which is, when any waggon-man loses his horse, the other waggon-men go a Gait (a journey to the Staith) for the poor sufferer, which is a little out of their profits, and purchase him another horse.[37]

In certain conditions rather unusual arrangements came into being whereby the goods would be carried part way by canal, part way by tram, again by canal, possibly even finishing their journey by tramway. Now shareholders in waterways were generally believed, not altogether fallaciously, to be making a lot of money from the vast dividends accruing to them. Regrettably, quite a number were merely 'daddy's sons' – no longer the men who had taken the risk of placing their money in the new-fangled canal in the first place. They were content to sit back, grab all they could and hang on to what they had got. If, however, your father had not been one of those sagacious gentlemen, might it not seem reasonable to seek an alternative way of making money? Whatever the cause there was discontent and to overcome this plans were being mooted, initially for the extension of horse-drawn tramways and with increasing vigour for the use of the 'tea kettle'.

'In speaking of the intervention of seven miles of canal, my readers will wonder why a communication of the roads has not been effected, to avoid the tedious process of unloading the trams, loading into a boat, unloading to a wharf, and loading into the trams again. They will be still more surprised when I inform them, that these seven miles of canal do not in effect take us an iota nearer market; and that by making four miles of new tramroad, not only all superfluous expense would be avoided, with loss of breakage, delay, &c., but also (in addition to the saving of seven miles of canal) four miles of the present tramroad would be avoided, by cutting off that circuitous portion of it which extends from the canal to the point where the junction would be effected. The sum requisite to form and complete these four miles of new road would not exceed £6,000 . . . It will hardly be believed, that with all the energy which those fine and opulent counties possess, no attempt has yet been made to effect so desirable an object. The importance of the undertaking is felt and acknowledged; the sums requisite for its completion are at hand; and I hardly know whether I am justified in noticing the apprehensions which seem to prevail, that from a narrow-minded and selfish policy, the proprietors of the canal (who would thus suffer a partial diminution of their profits) would be induced to give such opposition to the measure, as would be fatal to a bill introduced into Parliament for that purpose.'[36]

There we have an example of the tramway engineer, boat owner and mining expert; then we have the canal engineer Josiah Jessop who recommended a tramway to connect the Cromford Canal with the Peak Forest Canal. In October 1810, John Rennie put forward a quotation for a roughly equivalent canal line which would cost £650,000. By 1824, when costs had risen, Jessop's figure greatly undercut this.

'Having completed the survey of the proposed Railway from Cromford to the Peak Forest Canal at Whaley, and prepared the Plans, necessary to enable you to proceed to Parliament in the next Session; I now submit to you the Estimate for a double line . . . £149,206. 16. 8. The comparatively small expense of forming Railways will be a cause of extending our resources and finding new channels for capital and industry, that would forever have been neglected, if there were only the more expensive modes of Roads or Canals to resort to; the first being expensive in the carriage – the latter in the execution.'

Then we have the dissatisfied customers. Dissatisfied in this particular instance with the Bridgewater Canal and the Mersey & Irwell Navigation, for a number of well-argued reasons. In this context it must be remembered that the Victorians abhorred anything that smacked of monopoly. These two navigations were both owned by the Trustees to the Duke of Bridgewater's Estates, whose Secretary, Captain James Bradshaw, seems to have been a rather undiplomatic, heavy-handed, gentleman. Joseph Sandars represented a number of citizens who wanted to open a railroad between Liverpool and Manchester and in his 'letter' (actually a pamphlet)[38] to the members of both houses of Parliament appealing for this, he firstly drew attention to the manner in which the rates were compiled for carriage on the Duke of Bridgewater's Canal. In the authorising Act of Parliament for the waterway the flat canal toll rate was to be 2*s.* 6*d.* (12½p), this to include warehousing for a reasonable period. After the digging of the Trent & Mersey Canal from Runcorn to Preston Brook, which the Duke of Bridgewater undertook on behalf of the Trent & Mersey Canal Company's shareholders, His Grace was to receive the tonnage dues of 6*d.* (2½p) per ton on all goods which were destined to enter the Trent & Mersey Canal at Preston Brook after traversing his waterway. This was charged on *all* goods whithersoever they were bound, thus raising the tonnage duty between Liverpool and Manchester from 2*s.* 6*d.* to 3*s.* (15p). His Grace then found it necessary to build a reservoir at Runcorn wherein craft could lie until they could pass the locks, which

> he pretended to contruct for the convenience and despatch of his own vessels; but he had had the kindness to permit those of other carriers, on the condition that they pay him 1*s.* per ton for the privilege, and this he collects on goods passing

along his own Canal as well as on those destined for the Trent and Mersey. By this second device he raised his tonnage dues to 4*s*. per ton . . . the third and last device, for adding 1*s*. 2*d*. more to the tonnage dues, remains to be pointed out. At the time the Rochdale Canal Company obtained their Act for cutting a line from Rochdale to Manchester, His Grace obtained permission to make the Lock which now connects his Canal with the Rochdale; and for this he was empowered to levy 1*s*. 2*d*. on all goods which passed his Lock, for the avowed purpose of indemnifying himself from the loss which his warehouse property might sustain by the junction in question; but, in consideration of this, he was bound to find warehouse room, gratis, for a certain limited time. It will scarcely be believed, that instead of all this he exacts the 1*s*. 2*d*. per ton on all goods which are carried between Liverpool and Manchester, whether they pass the Junction Lock or not. Thus, you will observe, his Grace, by one manoeuvre or another, contrives to exact 5*s*. 2*d*. per ton on all goods that are navigated on his line, although it is perfectly clear that the Legislature never intended that he should have more than 2*s*. 6*d*. . . . There yet remains another exaction on the part of the Duke to be pointed out. All goods which pass up from Liverpool to Runcorn, to enter the Trent and Mersey navigation, pay above twice the freight which they ought, owing to the Trustees of the Duke having monopolised nearly the whole of the land and warehouses at Runcorn. At the present moment, they charge 5*s*. per ton on grain, while bye-carriers, in abundance, would be found to contract at 2*s*. But, if a bye-carrier arrive, the owner of the goods, who thought to save a few shillings a ton, finds that he had not only to pay the 2*s*. but also the 5*s*.; for say the Trustees, what will you do with the goods? Ours are the only warehouses, and your goods shall not be landed without your paying us as much as if we had carried them ourselves. Pay us the 5*s*. and your goods shall be received and forwarded.

Sandars then goes on to quote various examples of rates to prove his arguments. Perhaps these could have been overcome in the long term, but

> the necessity depends on the fact, that the present means of conveying goods are not adequate to discharge the business of the port, and that another line of conveyance has become requisite. It is requisite to give due facility to trade, and without it there will never be any permanent reduction in the rates of carriage, and no security against future combinations to advance them. It is quite clear that, if another line of conveyance is to be adopted, it must be a Rail-road, for another Canal could not be formed, because the existing establishment have engrossed all the water.

Here we have the appeal to the moral sentiments of the Victorians as well as a statement. Delays did occur, eight to ten days for the goods to travel from shipper to customer, eight to ten days to cover 50 miles. Perhaps now we do not consider this unreasonable but in 1824 it was. Insurance premiums had crept up due to petty pilfering of goods *en route* and Sandars stresses the case of spinners and dealers who are obliged to cart cotton on the high road, paying four times the price they would be charged by the proposed rail-road. He was not, however, over-optimistic when he quoted that, 'A Rail-road will enable passengers to travel between Liverpool and Manchester at the rate of 10 to 12 miles an hour, at one-half the price they pay now. A Rail-road will prevent any future combination to raise freights, and it would make the Trustees of his Grace of Bridgewater as anxious to let and sell warehouses and land, as they have been to grasp and retain them.'

Earlier the word schizophrenic was used to describe the thinking in Great Britain at the time of the coming of the railroads. In effect the country was divided into a pro-railroad bloc, as typified by Sandars, and an anti-railroad bloc, as typified by the writer in *John Bull* as late as 1834:

> To those who remember – and the stretch of recollection is not very great – the fatal year 1825, in which millions of property, hundreds of individuals, and thousands of characters and reputations were lost the present rage for railroad speculations must appear beyond measure marvellous. Companies spring up like daisies in the spring, and thousands

> and hundreds of thousands are nominally subscribed, before an engineer is even engaged to look at the line of country which it is proposed to destroy, and deposits to a vast amount are paid into counting-houses hired for the purpose, even before clerks can be engaged to take in the money and the subscribers . . . The commonest observation, a moment's thought, must show, that of these mad-cap undertakings, more than half, aye, perhaps two-thirds, must fail, and involve all the dupes of the prospectuses – for many have only got to that – before they know where they are. Those people who judge by the success of the Manchester and Liverpool railroad, and take it as a criterion for similar speculations, are dunces and blockheads. The distance from Liverpool to Manchester is about thirty-seven miles; the rail-road between those places was the first actually used upon a great scale; the intercourse between those towns is constant and general, the traffic great, and suitable to the speculation. It has hitherto answered; but it will be found upon a reference to the accounts of this undertaking, that its great source of profit had been the transit of passengers, a considerable number adopting it for convenience, but a very great many more from curiosity; and yet even this Eldorado had been proved unprofitable . . .

Undeniably, the writer was worried, not without reason:

> Does anybody mean to say, that decent people, passengers who would use their own carriages, and are accustomed to their own comforts, would consent to be hurried along through the air upon a railroad, from which, had a lazy schoolboy left a marble, or a wicked one a stone, they would be pitched off their perilous track, into the valley beneath; or is it to be imagined that women, who may like the fun of being whirled away on a party of pleasure for an hour, to see a sight, would endure the fatigue, and misery, and danger, not only to themselves, but their children and families, of being dragged through the air at the rate of twenty miles an hour, all their lives being at the mercy of a tin pipe, or a copper boiler, or the accidental dropping of a pebble on the line of way?

Then the writer passes to – oh, shades of a motorway! – the effect of railways upon the countryside.

> We, however, go farther, and denounce the mania as destructive of the country in a thousand particulars – the whole face of the kingdom is to be tatooed with these odious deformities; huge mounds are to intersect our beautiful valleys; the noise and stench of locomotive steam-engines are to disturb the quietude of the peasant, the farmer, and the gentleman; and the roaring of bullocks, the bleating of sheep, and the grunting of pigs are to keep up one continued uproar through the night, along the lines of these most dangerous and disfiguring abominations.

After dealing with accidents, not many, but he found a few, the writer sums up how he anticipated the whole matter would end.

> Railroads, without the slightest permanent advantage to the subscribers, or the public in general, will, in their efforts to gain ground, do incalculable mischief. If they succeed, they will give an unnatural impetus to society, destroy all the relations which exist between man and man, overthrow all mercantile regulations, overturn the metropolitan markets, drain the provinces of their resources and create, at the peril of human life, all sorts of confusion and distress. If they fail, nothing will be left but the hideous memorials of public folly, not cured by the exposure of Poyais bonds, Greek loans, and all the mining, and canalling, and other speculations of 1825, in the shape of ruinous, rotting mounds, the objects at once of the disgust and ridicule of those who have sense to appreciate, and prudence to preserve, the order of things as it exists, in perhaps the highest state of civilisation England has yet known.

It is at this point apposite to consider the history of one canal and the effects of the railway upon it. The Horncastle Canal, 11 miles long, extended from the junction of the Rivers Bain and Waring in Horncastle to the Whitham Navigation at Tattershall, giving an outlet for this inland town to Lincoln and the port of Boston. The capital permitted under its act of 1792 was £15,000, with powers to borrow £10,000. A so-called 'mania' navigation, like many it ran out of money and was forced to obtain a further act in 1800 to raise a further £20,000, the rates on goods being adjusted. The work then proceeded very fast and the canal was opened on Friday, 17 September 1802, with the usual jubilations. Surprisingly, traffic flourished and the population of Horncastle doubled over the 50 years from 1801 to 1851, but after the completion of their main line in 1852 the Great Northern Railway sought additional sources of income. The Horncastle Railway Bill met with opposition from the navigation company, some landowners and tradesmen but was duly passed in 1854.

> Horncastle and Kirkstead Railway. We are informed that during the past week the levels of the proposed line of railway from this town to the Kirkstead station of the Great Northern Co. have been taken. The line . . . will be contracted at a very moderate cost – less than £40,000 and it is believed it will obtain considerable traffic. Indeed, it is evident that if the present tedious and expensive means of transit, viz. the canal, can pay an average of 7 per cent. on the capital out of the carriage of heavy goods alone, a railway deriving its revenue from passengers, heavy goods, light goods, cattle – etc., etc., must be an equally profitable undertaking. . . . It is an important feature, that with the exception of a few of the canal shareholders, this project has the unanimous approval and support of the trading and agricultural interests.[39]

The railway was duly completed and opened a year later. The canal, however, did not go down without a fight, although they could not do very much. 'We hear that the dues on the navigation have been reduced very considerably in order to place them on a level with the charges of the railway, so that the public will reap the advantage of the competition.'[40] They were, one suspects, unlucky, when in December 1855: 'The late frost has completely suspended the traffic upon the Horncastle Navigation: if it had not been for the railway, the town and neighbourhood would have been in a deplorable state for coal, the stock in the merchant's yards being very low.'[41] Although surely the coal traders, had they not had the railway to rely upon, would have had the foresight to stock up with coals? The end came in 1871 when the navigation was finally closed. It is salutary to reflect however that the population of Horncastle has declined since 1851, for although the railways brought many conveniences they also brought cheap, factory-made goods and encouraged the depopulation of the countryside. Appositely enough the canal basin at Horncastle has almost disappeared and the dry dock is now the town's swimming pool!

It must not be thought that the coming of the Liverpool & Manchester Railroad in 1825 automatically ended the life of all canals, for even as late as 1833 *some* canal shares still maintained their high values.[9]

Canal	*Amount of share*	*Average cost per share*	*Price per share*	*Dividend p.a.*
Barnsley	£160	£217	£290	£14
Trent & Mersey (¼ share)	£50	—	£640	£37. 10
Coventry	£100	—	£600	£32
Erewash	£100	£750	£705	£47
Forth & Clyde	£100	£172. 13. 4	£290	£25
Grand Junction	£100	£224. 10. 0	£245	£12
Staffs & Worcs	£140	£140	£610	£34
Leicester & Northampton	£100	£83. 10. 0	£80	£4
Swansea	£100	£180	£220	£12
Ashby	£100	£113	£74	£4
Derby	£100	£110	£117	£6
Ivel & Ouse	£100	£100	£115. 10	£5
Portsmouth & Arundel	£50	£50	£10	—
Macclesfield	£100	£100	£50	—
Kensington	£100	£100	£10	—
Grand Western	£100	£100	£21	—
Basingstoke	£100	—	£5. 5. 0	—

True, these figures were not as good as they had been. In 1821 a £200 share in the Trent & Mersey Canal cost, on the open market, £1,750 and by 1824, £2,250. The Coventry Canal, the shares in which were nominally of £100 value, stood at £970 in 1821 and £1,350 in 1824, but by 3 April 1840 the Coventry was declining, down to £218 buying price just as the Trent & Mersey had fallen to £597 10*s*. 0*d*.

Ominously, as the railway companies grew in strength so some of the most vital connecting canals sold out to their competitors, including in 1846 the Trent & Mersey and the Birmingham Canal Navigations, in 1847 the Shropshire Union and the Stainforth & Keadby and in 1852 the Kennet & Avon Canal. By 1890 railway companies owned the following mileage of canals:

Company	*Mileage owned*
Great Western Railway	258
London & North Western	488
Midland Railway	50
Manchester, Sheffield & Lincolnshire	180½
North Staffs Railway	121
Caledonian Railway	60

In theory, it is true, tolls were tied but as we have shown drawbacks represented an essential factor if canals were to remain vital arteries of trade over long distances. For example, from London to Liverpool there were three possible routes:

Route 1	*via* Braunston Birmingham Autherley Eastham	244 miles 252 locks
Route 2	*via* Braunston Hawkesbury Fazeley Preston Brook	256 miles 191 locks
Route 3	*via* Braunston Napton Heywood Runcorn	267 miles 283 locks

Each one of these routes had a portion which was owned by a railway company who were not obliged to give drawbacks, although it is true they could and would, if it gave them a chance to snaffle traffics from under the eyes of rival railway companies! And again, it was illogical for railways, unless they could hope to be totally rid of the canal with its concomitant maintenance costs – an almost impossible proceeding – to eliminate all the traffic. But, in turn, they were unlikely to want to encourage traffic on a canal if it meant their relatively expensive locomotives and wagons lying idle.

By 1841 to send one ton from Liverpool or Manchester to London, meant tolls and carriers' charges which worked out at disproportionate rates:

Canal	*Miles*	*Toll* £	*s.*	*d.*	
Duke of Bridgewaters	25		1.	0	(5p)
Trent & Mersey	67		2.	9½	(14p)
Coventry (Fradley)	5½			2¾	(1p)
Birmingham (Fazeley)	5½			5½	(2p)
Coventry	21¼			11	(4½p)
Oxford	23⅝		2.	11	(14½p)
Grand Junction	101		4.	2½	(21p)
Total	248⅞		12.	6¼	(62½p)
Carriers' Charges		£2.	2.	5½	(£2. 12)

In 1947, immediately prior to nationalisation, the discrepancies in the rates became even more interesting.

Canal	*Mileage*	*Rate per ton* £	*s.*	*d.*	*Owner*
Ashton	15		11.	1.	Railway
Derby	14½		10.	0.	Private
Grand Union (Brentford–Braunston)	93	£2.	0.	0.	Private
Kennet & Avon	86½	£5.	16.	8.	Railway
Leeds & Liverpool (inc. locks)	127	£2.	1.	0.	Private

Canal	*Mileage*	*Rate per ton* £ s. d.	*Owner*
Oxford			
(Oxford–Hawkesbury)	77	£2. 10. 8.	Private
plus locks		5. 8.*	Private
Stratford	25½	£1. 17. 4.	Railway
Worcester & Birmingham	30	£1. 12. 0.	Private
plus towage through tunnels		5. 0.	
* minimum.			

We have already shown the basic reasons why coal traffic has declined on waterways. The direct effect on a waterway is shown in the gradual dwindling of the number of canal carriers. The old Grand Union Canal Carrying Company carried many thousands of tons to various factories in the London area, including Nash Paper Mills, John Dickinson & Sons, Colne Valley Sewage Works. These traffics were later carried by the British Transport Waterways in declining quantities until 1963. There were then some attempts by private carriers to keep coal moving, latterly greatly aided by the Ashby Canal Preservation Society, especially the traffic to Dickinson's Paper Mill at Croxley. It takes, we understand, one ton of coal to make one ton of paper and like all factories of any size Dickinson did not require fast delivery but a consistent flow, which circumstances are, of course, ideal for waterways. Sadly, the various carriers that were engaged in this trade were bedevilled by a number of small, niggling, occurrences – mechanical trouble with boats, the conveyor being out of action, difficulties with arranging a loading wharf and, indeed, obtaining the supplies of coal which had to be brought by road from the collieries to this wharf. With the present lack of dredging loads had to be restricted and profits, without a suitable back-load, became very near non-existent. The final runs were carried by a consortium of 'number ones' (owner-steerers), who mostly carried in their spare time.

The Ovaltine Company had a fleet of immaculately turned-out boats carrying to their works at Kings Langley but the craft were sold when the works changed over to oil-firing. S. E. Barlow, coal contractors of Tamworth, sold their fleet more or less contemporaneously, then Harvey-Taylor of Aylesbury followed by another coal carrier, Gilbert Bros. of Bedworth, operating as the Warwickshire Coal Carrying Company. These were only a few of the number who were doomed to die in the early 1950s. The last independent company to and from the coalfields was Samuel Barlow of Braunston, some of whose boats were bought by Blue Line Canal Carriers, Ltd., and operated quietly up and down the Coventry, Oxford and Grand Union canals, albeit with only three pairs, until 1970 – operating from Atherstone, or thereabouts, according to loading facilities, to the 'Jam 'ole' at Southall. It was a rather sad coincidence that a matter of months before the cessation of this traffic, Michael Street, the then owner of the Blue Line Canal Carrying Company had just obtained a number of relatively new craft to replace some the life of which had expired. Almost simultaneously, Captain Ward took the last pair of Willow Wren Canal Transport Services boats 'dahn the smoke', thus to all intents and purposes ending commercial carrying of coal in any quantity from Birmingham to London.

A wide-boat canal, in so far as the locks could take barges measuring 66 ft by 15 ft was the Barnsley. The principal objects were 'to open a navigation from Sheffield and Rotherham to the other numerous and important towns of Yorkshire and Lancashire, and to bring into use the coal of Barnsley, Silkstone, and the neighbourhood'.[42] Receiving its authorising Act in 1793, the company was permitted to raise some £72,000 plus £200,000 by mortgage, but this did not suffice and a further sum of £43,200 was called for in 1808. This canal was, without question, of great advantage, both in so far as it paid good dividends to the shareholders and 'its execution has had the effect of introducing the coal worked in the latter place [Silkstone], into the London Market, where it holds a distinguished place among the Yorkshire Coals. The making of this canal has also been of incalculable advantage to the agriculturists in its vicinity, by the facility it gives to the introduction of Knottingley Lime.'[4] The canal proved to be very successful from the very beginning, and even the railway period which commenced in the 1840s did little to deter traffic. In fact the canal was financially attractive to the Undertakers (the title of the authorising body) of the Aire & Calder Navigation, who leased it in 1855, eventually purchasing it outright on 30 November 1875 at a cost of £48,800. In 1893 the section of the canal from the bottom of Barugh locks to the terminal at Barnby Basin was authorised to be abandoned, for the Silkstone pits were worked out. Despite this the coal traffic continued to prosper with the opening of new collieries between Barnsley and Wakefield.

During the 1940s the canal suffered three major stoppages. The first, for three weeks, from 27 October to 16 November 1942, the second of four months' duration, from 16 June to 18 October 1945 and the final and most disastrous stoppage, directly caused by a breach of the bank which occurred on 27 November 1946, resulted in the canal being closed to all traffic from 27 November 1946 to 1 May 1947. The obvious result of these stoppages was the breaking of contracts, and the resulting loss of trade and loss of livelihood to the bargees. Many barges, mainly of timber construction, were abandoned and left to rot. Most of the Rogerson's fleet, who were regular by-traders on the Barnsley Canal, were abandoned. After the re-opening of the canal, now under the authority of the British Transport Commission, there was little hope of succeeding in keeping it open, especially as officialdom had declared that the Barnsley had no future, so it was only a few years before the abandonment in 1953. Most of the length of the canal has now been filled in, and only a few isolated stretches remain to remind one that it was once a most successful canal, paying dividends at one time as high as 15 per cent.

It has already been shown that when waterways came on the scene Britain was still predominantly agricultural, both physically and mentally. Canals, and the quick transport they offered, invited the building of factories on their banks.

> The village of Etruria is wholly of the late Mr. Wedgwood's creation. It is locally situate within the township of Shelton, upon the turnpike-road from Newcastle to Leek (two miles distant from Newcastle) and consists of the mansion called Etruria Hall, now occupied by Mr. Francis Wedgwood, a partner in the firm of Josiah Wedgwood and Sons; the large

44 Strengthening of Armitage Tunnel due to mining in the area

MT 2430

manufactory, which occupied the western bank of the canal, two branches from which are carried into the manufactory itself, and a continuous street of about 120 workmen's dwellings adjacent, with an inn, and some houses of a better class, for farmers, clerks, and others.[22]

To serve these new factories people – men, women and children – were necessary and when Josiah Wedgwood was called before a Select Committee of the House of Commons in 1816 he no longer showed the interest in the pleasant things of life the family enjoyed during the building of the Trent & Mersey Canal. Instead by his own words and those of the Inspector, he exposed the true Victorian businessman's nature.

'Extensive manufactory in potteries, employs 387 persons, of whom 13 are under ten years and 103 between ten and eighteen. Irregular buildings, much scattered. Work-people now crowded, doors generally open, casements in windows which work-people may open. Work generally unhealthy but that part connected with applying the glaze is composed in part of white lead and if workmen careless in their method of living and dirty, very subject to disease, depends much upon care men take of themselves. Dippers work from 8 or 9 to about 5 in the afternoon with interval of half-hour for breakfast and 1 hour for dinner. Hours of work for other persons from half past six to six in the evening, with half hour for breakfast and an hour for dinner. The hours are shortened when they cannot see without candles except from 11th November until 3rd February in which period candles are burnt and they work from half past seven to six. Children work same time, the younger children employed in attending the men, assisting them in carrying their moulds. Wages of children paid to parents. Much too often these additional wages are employed by the parents not in giving the children improved clothing and food but in procuring spirituous liquor for themselves.'

'Then the Committee is to understand from you, that although you think that the curtailment of labour to children

45 A River Lea barge being launched, *c.* 1934

would be extremely beneficial, you object to an Act of Parliament, not for containing these clauses which would be beneficial, but you object to any legislative interference with the manufactories?'

'I have a strong opinion that, from all I know at present of manufactories in general, and certainly from all I know of my own, we had better be left alone . . .'

But what of the conditions elsewhere? Josiah Wedgwood was, within his own lights, a humane man. Thomas Gibson and George Bryan were witnesses before this same Select Committee during the investigation into the Duke of Bridgewater's coal mines at Worsley.

Both worked in coal mines since boys.
'What had you to do then?'
'Thrutching the basket and drawing. It is done by little boys; one draws the basket and the other pushes it behind. Very hard labour. Worked nearly nine hours a day regularly, sometimes twelve, have worked above thirteen. We used to go in at six in the morning, and took a bit of bread and cheese in our pocket and stopped two or three minutes; and some days nothing to eat at all – we were overburdened, I had only a mother, and she had nothing to give me, I was sometimes half starved. They still have only a bit of bread and cheese sometimes can't eat it all owing to the dust and damp and badness of air; sometimes as hot as an oven, have seen it so hot as to melt a candle.' A boy of eight got 3*d.* or 4*d.* a day. A man's wages divided into 8, at eight years gets one-eighth, at eleven two-eighths, at thirteen three-eighths, at fifteen four-eighths, at twenty man's wages, about 15*s.* in full employment, often not more than 10*s.* Out of this he has to get his tools and candles. Consumes four candles in nine hours work, in some places six, 6*d.* per pound and 24 candles to the pound. Beaten a score of times, both purs [a local term for a kick from a clog] and kicks and thumps. Many girls employed, do the same work as boys, till they reach 14 when they get wages of half a man, never get more, and continue at the same work

for many years. Both boys and girls fight, sometimes they are very loving with one another . . .'

At Bilston, a witness, again before this Select Committee, which was very thorough, stated that

'Besides my husband I have a boy who works on the band [breaking up the coal on the conveyor chain or belt] with my husband, a dirt carrier, at 2*s*. a day. We have also an apprentice, but we don't know his age nor he himself except what we guessen. He came from Manchester; a boatman picked him up upon the canal side and brought him with him to drive his horse, and he could not find his way back. He was about 10 or 11 years old, named William Butler, and he used to lie about and get burnt at the coke hearths. A woman brought him to me and asked if I would have him; and I pieced [bought] a pair of trousers and waistcoat for him, and a pair of shoes that were too small for one of my wenches I had tipped and put on his feet. He bound himself apprentice.'

Boats had to be loaded at coal mines, not always as you might think, by crane; there were other methods.

On the banks of the canals in Staffordshire are seen many girls engaged in loading the boats with coals. These girls are substantially, though coarsely clothed, and the head and neck more particularly protected from the cold. The work is laborious, but not beyond their strength. The clothing is obviously such that a girl cannot continue to wear it after going home. She therefore lays it aside, and washes herself, and puts on more agreeable clothing for the rest of the day . . . The returns from Staffordshire show the wages of the girls to be as follows:

Age	*Pay*
12 to 13	4*s*. to 4*s*. 6*d*.
13 to 14	4*s*. 6*d*. to 5*s*.
14 to 15	5*s*. to 6*s*. 6*d*.
15 to 16	6*s*. to 7*s*. 6*d*.
16 to 17	7*s*. to 8*s*. 6*d*.
17 to 18	7*s*. 6*d*. to 9*s*.

Sometimes, then as now, canals had other uses than those they were designed for, and perhaps the most horrific part of this report in 1852 is the casual acceptance by the victim of what was to happen.

> 'My father was a porter at the railway station; and he came home drunk when he got paid, one Friday night; and he took James and me, and he said he would take us to the canal and drown us. He told our step-mother to reach our shoes; she said "If you are going to drown them, you may as well leave their shoes for Johnny". He took us, and he threw me in; and I should have been drowned only for a boatman. There was two policemen on the bridge, with their lamps; they did not come to us then but they came to our house after we came home; and they said they came to see about those two children; it was not a proper time to be on the canal side that time in the morning (2 o'clock).'[43]

In Victorian times England and Wales were booming countries, we had great markets throughout the empire and there was a drive to get things done whatever the circumstances and despite regulations put out by the House of Commons. Every man upon whom the breath of civilisation had touched possessed the drive to do better, the fact that to do better meant trampling on the shoulders of other people was irrelevant. To make money there was trade. Canals brought the coal to make the goods and took away the goods to 'mek t'brass'. Conditions were, without doubt, appalling both in housing and nourishment. In the main men who came from the countryside did not easily adapt to city life in a hovel but equally because they lived in a hovel there was a drive to make matters better. There was no sick pay, apart from the subscription clubs, there was no National Health, you paid as you went, but there was no limit to how far a man could go in the manufacturing field.

> Wanted. A Person who is acquainted with the Manufactures of heavy Iron Goods, in various Forms, to Superintend a Work of that Kind, which is established at some Distance from Birmingham, very advantageously situated for Coal, Wood,

and the Exportation of its Produce; or, in any sober and industrious Person, with a Capital of One or Two Thousand Pounds, more or less, who has been brought up to the Iron-mongery Branches, and accustomed to ride Journies for Orders, &c. should want an Establishment, and incline to become a Partner for the chief Conduct thereof; he shall have a Share therein proportioned to any Capital which he may bring, besides an handsome Allowance from the Company for his Superintendance of the said Work.[44]

There is a saying in Essex, 'from muck to muck in three generations'. The dozen of little factories that sprang up in every industrial city, Leeds, Manchester, Birmingham, etc., during the nineteenth century are mainly doomed. A few have grown and enlarged themselves by swallowing up the small manufactories during trade recessions in the twentieth century, but the remaining small factories are out-dated.

One car factory worker, talking to us on the canal bank about housing and working conditions expressed the opinion that these have not altered much although they have been up-dated. Instead of a hovel there is a flat, perched 20 storeys up in the air; instead of the hot, humid, closed-in atmosphere of a working steam-powered factory there is a clinical, soulless, electrically-operated machine. He explained that he had to leave his flat at five o'clock in the morning to clock-on at six and then produce 1,000 items by clocking-off time at two o'clock. Then he drove home and had his dinner and watched the telly and so on, almost *ad infinitum*. He came to the canal bank to fish and enjoy the peace, little realising that where he sat there was once an industrious factory producing machines to make machines to make cars.

Given a reasonably short haul (not in excess of 20 miles), level countryside and a reasonably good turnpike road a road-carrier could compete on more or less equal terms with the canal, providing always that his load did not exceed five tons. For this reason parcel boats were not, in the main, successful ventures unless they could generate heavier traffics. To give an example, immediately prior to the First World War, Messrs Fellows, Morton & Clayton ran a scheduled boat from Ellesmere Port to

Wolverhampton, its purpose being to pick up goods of a reasonably light weight which were to be transported to destinations *en route*. If the first boat became full a second boat was immediately pressed into service. It was widely known that if the boat's captain could pick up a 'parcel' from a factory not hitherto served by the company Messrs Fellows, Morton & Clayton, hoping this would lead to further traffics, were happy to pay the captain the, then, magnificent sum of £5 by way of a bonus. The boat itself did not, and could not, pay but it was a service which they hoped would catch the interest of any of the factories which favoured road or rail transport. There was no reason why any responsible carrier, prepared to run his craft on a fixed timetable of departure should not run such a service as it was possible, even until the early 1900s, to make up a full load of two- five- or ten-ton lots; but on some canals restrictions hindered such operations. The Act of Parliament authorising the building of the Newcastle-under-Lyme Canal penalised boatmen with small loads, or empty boats; showing under the tonnage rates 'if less than a ton in a boat 6*d*. per mile', the normal rate being 1½*d*. per ton per mile. The River Ouse, Sussex, in its rates, showed 'No Vessel of less than Ten Tons to be passed through the Locks, without leave of the Proprietors', while the Peak Forest Canal were determined to both have their cake and eat it, 'No Boat under Fifteen Tons Burthen, when the Water does not, or under Ten Tons when the Water does, flow over the Weirs of any of the Locks shall pass through any Lock . . . unless the Navigator of such Boat shall pay Tonnage equal to Fifteen Tons or Ten Tons respectively as aforesaid.'

Most waterways had a restriction of some sort. It was, of course, reasonable to the extent that they did not wish to lose water; most waived the charge for a boat working back empty, but should it pick up even one parcel the rate was firmly enforced. Let us not, however, be misled into believing that in nine cases out of ten the would-be consignor in the early nineteenth century had any choice between road and canal. 'Roads in general, very middling, they look forward to the completion of the canal [Hereford & Gloucester] for very great improvement of the roads, not only by easing them of much heavy carriage but, for the supply of good mending materials which are now, in many situations, scarce and distant.'[45] From this extract it will be seen that the ill-fated Hereford & Gloucester Canal was regarded not only as a means of lowering the rates of transportation but, indeed, as a means of improving access to the interior of the country. Without the land drainage of today, aided in many places by the canals, there was little hope for the old-fashioned soft-surfaced roads when the winter rains came. If it rained, wagons, horses and men alike got mired down and if it was frosty a horse's leg might easily be broken. Even in November many roads could be impassable, as they were in 1770:

> By the very heavy and continued Rains, the Roads in many parts of Staffordshire and in this Country are rendered extremely dangerous: We hear of many Accidents occasioned by the Floods in Derbyshire, &c. . . . Our Mails have arrived several Hours later than usual; and our Accounts from the Counties of Worcester, Gloucester and Hereford are of a similar Tendency. Last Thursday as Mr. Robert Willis of Tamworth, was passing the new Bridge at that Place, the Rapidity of the Flood carried away one of the Arches, by which Accident he was unfortunately drowned. At Wrighten, near Coventry, the Bridge has been broke down by the Floods, which detains a Number of Carriages and Passengers at that Place. Yesterday between Beacon's End and Tamworth, a Waggon was overturned, when two of the Horses fell into a Pit and were drowned.[46]

These conditions were, if not accepted, at least tolerated, and it would seem difficult to visualise that matters could be otherwise.

> Though this useful set of Men, the Farmers, will undoubtedly reap a Proportion of Advantages from the Execution of this beneficial Scheme, they are far from being satisfied, and seem to reflect upon it with many Doubts and Fears. Custom, indeed, and Occupation in Life, cast a wonderful Influence on the Opinions of all Mankind; it is therefore by no means surprizing that men, whose Forefathers, for Ages, have been

inured to rugged and deep Roads, to wade after their Beasts of Burden up to the Knees in Mire, to see their loaded Waggons stick fast in Dirt; Men, who from their interior, inland Situation, are almost totally unacquainted with all Objects of Navigation, it is by no means strange that People, so unaccustomed, should consider an Attempt to introduce a navigable Canal up to the Town of Chesterfield, and within the Air of the Peak-Mountains, with alarming Ideas, with Suspicion and Amazement.[47]

As time went on various attempts were made to introduce steam transport on the new turnpikes. These turnpikes were, in the main, an excellent arrangement, for quite simply a section of road was improved by having a good surface made; the right to levy tolls was given to the proprietors. There were some abuses of this system, for the right of collection of tolls was normally put up to auction, as we find in a notice of 1805.

Notice is hereby given, that the Tolls arising at the several Toll-gates erected upon the Turnpike-roads leading from Lutterworth to Hinckley, in the County of Leicester . . . will be let by Auction to the best Bidder, at the House of William Tomlinson, known by the Sign of the Bull's Head Inn, in Hinckley . . . on Tuesday, the 16th Day of April next, between the Hours of Eleven and Five o'clock of the same Day, in the Manner directed by an Act passed in the Thirteenth Year of the Reign of his present Majesty, King George the Third, for regulating the Turnpike Roads, which Tolls were let for and produced the last Year the following Sums, namely.

	£	*s.*	*d.*
The Bitteswell Toll-gate, for	148.	0.	0.
The High Croft Toll-gate, for	134.	0.	0.
And the Burbage Side-gate and the Lickett-Lane Side-gate, together by Computation, produced	28.	0.	0.

Over and above the Expences of collecting the same, and will be put up at those respective Sums; and the highest Bidders or Takers of the same respectively, may enter upon

> the Tolls arising therefrom upon the 21st Day of April next . . . By order of the Commissioners, CHARLES JERVIS, Clerk.[48]

Very often a canal act granted the toll-free passage of road stone, thus aiding its competitor, the turnpike, for having only the carriers to pay, stone could be brought easily and economically to the nearest wharf.

It is interesting to compare accounts of roads. In 1768:

> '[the road] is for near twelve miles so narrow that a mouse cannot pass by any carriage: I saw a fellow creep under his waggon to assist men to lift, if possible, my chaise over a hedge. The ruts are of an incredible depth . . . The trees everywhere overgrow the road, so that it is totally impervious to the sun, except at a few places; and to add to all the infamous circumstances which occur to plague a traveller, I must not forget the eternally meeting with chalk waggons, themselves frequently stuck fast, till a collection of them are in the same situation and twenty or thirty horses may be tacked to each to draw them out one by one.'[49]

In 1832:

> The Roads of Nottinghamshire have been greatly improved during the last ten years . . . but in some of the lanes and bye-roads the traveller still has to contend with deep sand and in some places with an adhesive clay, which latter he finds as fatiguing in wet weather, as the former is in dry . . . The road from Nottingham to Loughborough was almost impassable until 1738 . . . The road from Retford to Worksop, has now a good 'macadamised' bottom though a few years ago it was the heaviest sandy road in the county, except that from Worksop to Warsop, in which Laird says he was three hours in travelling 8 miles in a post-chaise.[50]

The earlier steam-engines were more or less prohibited by the high charges placed upon them by the owners of the turnpikes, and even traction engines, with their enormous capacity, still presented a weight problem which often meant they had to be diverted many miles out of their direct route to get from one point to another – a diversion often caused by canal bridges! A Fowler Big Lion of 1890 could handle 40 tons but it weighed $16\frac{3}{4}$ tons; the Super Lion, derisively rated at 10 nominal h.p., could pull 120 tons, but weighed $10\frac{1}{2}$ tons without its load of coals and water.

An entirely different competitor was the coasting vessel, certainly for long-distance traffics, Liverpool to London, Hull to London and the like. As long as sailing craft were utilised, with their attendant long delays, a state of easy co-existence was reached between them and canal boats, but as the steam coaster became more reliable so it was axiomatic that it would draw traffic from the canals, the very traffic that canals were designed to cope with – bulk. All things considered, from Nottingham to London a 100-ton load would require four men, plus their wives or lads, four boats, four horses and five days on inland water; a steam coaster, to which 100 tons was only a part load, would require four men only and could probably complete the journey in 48 hours. The railway, of course, could outdo both in terms of speed, but could not undercut either until comparatively recent times.

Motorways can be regarded as a political pawn – Adolf Hitler used them in Germany and Benito Mussolini in Italy as a means of absorbing unemployment. In lieu of our present-day dole, the men employed on such works were fed, watered and housed and their families looked after by means of a grant, which could be encashed at certain stores which, at least, obviated the present trend of spending the dole, or the Family Allowance, at the public house. In Britain the cost of motorways is very high and following, as we usually do, the American pattern, there is very little doubt that extra charges will be made, if not tomorrow then the next day, for the use of them, although whether by tolls or a licence differential remains to be seen. Boats and barges have a minimum life of 15 years, they consume diesel fuel at the rate of approximately half a pint per horse power per hour. Twenty horse power is adequate to move a pair of laden (55 tons) narrow boats along the midlands canals and 120 h.p. is adequate to move 450 tons along the wide canals of the north. A lorry capable of pulling its own weight and carrying ten tons has an engine of about 60 h.p. and consumes fuel at between four and eight miles per gallon. This lorry in first-class motorway-hammering trim has a life of about five years. The barge will use an occasional drop of lubricating oil, a small amount of rope for its fenders, coal for its stove and little else. A lorry requires tyres, anti-freeze, windscreen-wipers, repairs and maintenance. A boat's engine will have a life of five years, a lorry's two. By deliberately eliminating canal traffic and encouraging that on the roads, governments are, in their own way, reducing the likelihood of unemployment in the motor industry and ensuring a greater income by way of motor-taxes, insurances and fuel taxes than could be obtained from boats and barges. The fact that using canal transport could reduce the cost of living is, in the light of the above, entirely irrelevant.

Niggles have always abounded on canals. Niggles which have meant reduced payloads, canals going by the least direct route due to landowners' influence and niggles which, on occasion, have stopped canals from even happening. The Blythe Navigation on the Fens relied upon the harbour at Southwold for its outlet and when, due in part to inefficiency, in part to personal antagonisms and in part to parsimony the harbour eventually silted up, the navigation was finished. The North Walsham & Dilham Canal in Norfolk was built under an Act of Parliament of 1812, there being nothing unusual at the time of building except that the rate 'for all passengers in Boats or Barges . . . 1*d.* per mile each' was specified. In 1893, the general state of decay in the neighbourhood was noted by a pleasure craft hirer. 'They had the greatest difficulty in getting above Dilham; the river was nearly dry and they were told: "There's only been one wherry up since Christmas. Are you going up to Antingham? . . . that's a long way. I shouldn't like it. Right, you go!" . . . signs of decaying trade – a deserted granary, a decaying landing stage . . . a ruined corn mill. It was half a century and more since Thomas Fowell Buxton, moving to his new home at Northrepps Hall, had his furniture from London by sea to this very quay.'[51] Lewis Edwards, Honorary Secretary of the East Anglian Waterways Association, has always been in the forefront of attempts to restore this canal to a navigable state, since it was last used by hire boats in 1934. To extinguish the navigation rights would involve a good deal of money in compensation

– the extinguishment of the millers' rights alone would be an expensive business. The canal runs through very attractive countryside and the scene of dereliction at Horning Lock gives no real clue to its possibilities. For a dinghy there is water through delightful woods up to the next lock, and the scenery is quite enchanting as it is typical of the North Norfolk country which is hardly ever seen by the Broadland visitor. The whole waterway should, clearly, be transferred to a Trust, perhaps similar to that on the Lower Avon Navigation.

Another form of niggle which has affected a number of navigations has been that occasioned by business rivalry, the most blatant being reported on the Glamorganshire Canal when, following a dispute between two local ironmasters in 1794, the offended one wrote: 'I understand that Mr. Crawshay has broke the locks and let off all the water from under the boats loaded with our iron . . .' Not really cricket, that! Such a trend was felt in the very last days of independent carrying on narrow canals when a lock would be drawn off (emptied) in the face of an oncoming pair of boats, a proceeding which, apart from wasting water, contributed towards delays in traffic. Many were the brawls, even in the 1950s, at bottle-necks on the canal system – Farmer's Bridge, Camp Hill and Tipton. Is it a curious hangover of this that in the amenity age three 20-ft boats will work a flight of locks, even the 'thirty' at Tardebigge, independently rather than together? The third in such a queue surely vows never to go cruising again.

In 1890 a not very enthusiastic writer on waterways stated categorically that:

> It would be the idlest of idle dreams to expect that the canal system of this or any other country, as originally constructed, can be resuscitated, or even temporarily galvanised into activity, in competition with railways. Canals as they were built a century ago have no longer any function to fulfil that is worthy of serious consideration. Their mission is ended; their use is an anachronism. They do not provide the means of cheaper transport, and they have no other advantage to offer to the trader that would be a sufficient equivalent for the tedium of their transport.[52]

And it is curious, but true, that even in the First World War canals were worked to nothing like their capacity, so much so that in 1919 the Canal Control Committee of the Board of Trade made a rather desperate appeal for their greater use.

> The principal method of transporting goods in this country is by railway. The railways have already rendered a great service to the country during the War. The demands which have been made upon them have increased, while the staff which is available for their administration is reduced . . . There is a limit to the work which the railways can do for the Country, and, consequently, it is of the utmost importance that other methods of conveyance of heavy goods, and indeed, of all classes of commodities, should be utilised to the same extent during the War. Motor traffic is not available to the same extent as it was. Certain facilities, however, do exist for transporting goods by canals and inland waterways . . . It is desirable that senders of heavy goods, manufacturers, and traders throughout the Country should render every possible assistance to the Government in dealing with the difficulties which have arisen in the transport of goods. The necessity is becoming more urgent as the War proceeds, and it will, therefore, be a patriotic step on behalf of all who can do so, if they will use the inland waterways of the Country, when they are suitable for the transport of their commodities between one point and another.[53]

The same pattern arose in the Second World War and the comments of Mr E. J. Woolley, when addressing the Grand Union Canal Company and presenting their Annual Report for the year 1939, make interesting reading.

> 'When we met here a year ago it was under the passing shadow of a great crisis. Unhappily to-day we meet under the black clouds of what may prove to be the most decisive war in the history of the Empire, and I feel certain that you will appreciate that my remarks must be tempered with everything which that implies . . . But I have no hesitation in saying that if war had not broken out the accounts would

have presented an even better appearance, and the board would have probably recommended some payment being made to the Preference stockholders. As it is we still find it impossible to pay a dividend . . . It is a lamentable fact to have to relate that for the last four months of the year there was a decrease of no less than 131,515 tons compared with the same period last year, and the fall in tonnage revenue amounted to £6,145, so that even if the trade had continued on the same level after the outbreak of war we should have still shown an improvement.

'As you are well aware, a considerable portion of our traffic arrived in London from overseas. The diversion of ships to other ports must therefore have a direct bearing on our results, not only as regards the traffic carried on our waterways, but also on our dock receipts, for we lose the dock dues and quay charges, and our stevedoring and wharfage company at the same time loses revenue.'

After giving the reasons for the decline in tonnage – primarily the now non-existent four steamers per week that had previously operated out of German ports, coal no longer being brought by coast-wise steamers, and the drop in revenue from timber traffics, Mr Woolley went on to say:

'On the other side of the picture it is pleasing to be able to record that maintenance shows a fall of £2,620, although the company has been faced with heavy expenditure in connexion with I.R.A. outrages earlier in the year, and our troubles have been further accentuated by a serious burst in the canal banks at Weedon on October 18th. This was due to the exceptionally heavy floods, which I am informed were the most serious in living memory, both by reason of the large area covered and the enormous volume of water rising in a very short space of time. This cost us £1,180 in repairs. But unfortunately it necessitated closing this part of the canal for 10 days, and in consequence some thousands of tons of traffic were lost. It is to be hoped that both these items may be considered as non-recurring expenditure.'

After covering the revenues from the estates, stevedoring and wharfage divisions, he continued into a report on the difficulties found by the Grand Union Canal Carrying Company which

> '. . . has experienced a year of somewhat mixed fortunes; but in spite of the extremely difficult conditions under which it has operated the company had made a trading profit of £1,779, as against a loss of £3,349 in 1938. Once again throughout the year the company has suffered from the lack of skilled boatmen, which seriously handicapped its operations, and in consequence many thousands of tons of goods were lost. Although the actual traffic booked by the company amounted to 211,959 tons, an increase of 16,178 tons over last year, it was only able to carry 166,044 tons in its own boats, the balance of 45,915 tons being given to other carriers operating on our canal. The tolls on traffic booked by our carrying company amounted to £22,309, compared with £17,466 in 1938.'

As a result of the war some new traffics were generated on waterways, including milk powder from Brentford to the Ministry of Food Depot at Blaby and flour, again for a Ministry of Food Depot, from Brentford to Tring. A very popular traffic was sugar, again from Brentford and destined for the Ministry of Food Depot at Marsworth – popular with boatmen for obvious reasons! But, overall, the same picture arises of a run-down of the canals, maintenance decreasing and traffics with it; in effect the preparation for the *coup de grâce* given by nationalisation.

9 *At Your Leisure*

> Mention has been made previously of the Buxworth Basin Restoration Scheme so valiantly pursued by the Inland Waterways Protection Society. Members will be interested to hear that the first 100 yards of this restoration received an award under the European Conservation Year 1970 Countryside Awards Scheme. The award consists of a special designed bronze plaque and a signed certificate and was unveiled at a ceremony on the site at Buxworth on 1st November last. We congratulate the I.W.P.S. on their splendid effort and wish them every success with the restoration of the further lengths now in progress.[54]

This is one restoration scheme among the many either doing or mooted. The work on the southern end of the Stratford-upon-Avon Canal (Lapworth to Stratford), carried out under the aegis of the National Trust, has received commendable – if sometimes misleading – publicity. Misleading because although the work was well done the canal was still there with about 90 per cent of its works complete, and only negligence and lack of traffic had brought it to an almost unusable state. The publicity is also misleading in so far as that upon reading it one imagines

46 Dukinfield, Ashton Canal

47 Pleasure boat at Shipley, Leeds & Liverpool Canal

48 No. 4 Lock, Ardrishaig, Crinan Canal, on the day of the Clyde Cruising Club's Tobermory Race

49 Decked tar boat leaving the 'ampton twenty-one, September 1964

that we have a new canal, and one that requires no maintenance. Bitterly the manager regrets that this is not so. However, this canal restoration is almost solely for the benefit of pleasure craft. The extract at the beginning of this chapter relates to a rather different arrangement. The Inland Waterways Protection Society cannot be called a large organisation, but they are, without misusing the word, dedicated. The Buxworth (ex-Bugsworth) complex of basins at the head of the Peak Forest Canal fell out of use in 1953 and was steadily allowed to deteriorate. In 1964 its restoration was mooted and slowly but surely, after defeating various opposition groups from within the nationalised industries, councils and private bodies, some 300 yards of the waterway has been wholly restored. By wholly restored we should say that, unlike any other restoration, this waterway will have its original depth (5 ft 6 in.) of water and full width. Curiously there is a greater resistance to restoration – and hence lack of money – in the north than in the south.

Another radically different navigation work, inasmuch as it involved cutting new lengths of waterway, is the revival of navigation to the Upper (Warwickshire) Avon, which when completed will give the Stratford-upon-Avon an outlet to the sea, whereupon it will be possible for pleasure craft to make a round trip from Worcester down the river Severn to Tewkesbury, thence via the Avon to Stratford, the Stratford Canal to Kings Norton and the Worcester & Birmingham Canal back to Worcester.

Another restoration mooted is that of the Ashton Canal in the Manchester area. Being taken under the wing of the Inland Waterways Association it will re-open, again for pleasure craft, the 'Cheshire Ring' route – Rochdale Canal, Ashton Canal, Bridgewater Canal, Trent & Mersey Canal, Macclesfield Canal.

So much for what is being done in Britain, in effect restoring the *status quo* to that of 1830. On the Continent, however, they consider a canal which can only take carrying craft of 300 tons to be hopelessly outmoded and are seriously talking in terms of 1,000-ton (Imperial measure) barges. So whence comes the difference and why are pleasure craft the mainstay of our waterways?

There are three possible treatments for a waterway which, like those in Britain, has reached the end of its life in its present form.

The first is to close it. There are a number of ways of closing a canal which we shall return to later. Any such proposal will meet with a certain degree of opposition.

The second is to state that such waterways will become cruiseways, indicating that they are suitable for pleasure craft and maintenance standards can be reduced to that level. Commercial traffic is tacitly discouraged.

The third possibility is to enlarge and modernise them – expensive, but not more so than building a motorway.

Pleasure craft have been in existence since the inception of waterways. The Duke of Bridgewater was quite prepared to board the better class of barge used on his navigation and inspect the works proceeding.

> 'From early impressions and recollection of youthful scenes and adventures, I feel a strong attachment to canal conveyances. I remember the delight with which I entered the Duke of Bridgewater's canal boat, to enjoy an excursion from Worsley Mills to Runcorn – I remember the noble Duke himself – I remember his dark brown coat with gold buttons, the handsome rows of black ribbon which ornamented the tie of his hair, his placid but cheerful countenance, his manly and dignified form and carriage – they will never be forgotten. I remember his handsome pleasure boat; I can almost fancy I still see the Duke, the Lord Gower, now Duke of Sutherland, and Mr. Bradshaw, standing on the forend of the vessel on a fine summer evening, enjoying the refreshing breezes excited by quick sailing.'[1]

The Duke set a precedent which was followed by Wedgwood and his friends on the Trent & Mersey Canal. From such joy rides, came, logically enough, the Directors' Inspection Launches. On the Chelmer & Blackwater Navigation, the directors of the canal company and their guests still make an annual inspection of the line, a custom which has survived from the early days of the navigation and had its counterpart on most of the other canals and inland waterways in Britain. Due to the

South Portal, Dudley Tunnel.

loving care bestowed on them a few such launches still remain in existence today, although they are perhaps akin to the Irishman's knife which had had three new handles and two new blades but was 'the same knife me grandfather gave me'. The *Lady Hatherton* of the Staffs & Worcs Canal had her old, but very elegant and serviceable, cabin placed on a new hull fairly recently, the old hull being sold later to become a residence, with a new cabin, for some 'way out' gentlemen. The British Waterways Board still owns an inspection launch, the *Water Rambler*, an ex-zoo bus, which meanders around the midland circuit, poking its nose into some of the lesser-known canals, but seemingly only during the summer months.

Passengers on packet boats, which as we have shown sprang up on a large number of canals, were in the main worthy citizens going about their business, but a few took trips on the water for pleasure and one or two were courteous enough to leave us notes of their experiences. In 1833 we have a description of a trip on the River Trent.

> If the Traveller be not wholly devoid of curiosity and taste, he is impelled to notice the local peculiarities . . . In the steamboat, however, where a number of fellow-travellers are congregated together; the social pic-nic not unusually becomes engrossing, and it is not an unfrequent occurrence for the beautiful and noble scenery on either side of the majestic Trent,
>
> who like some earthborn giant spreads,
> His thirsty arms along the indented meads
>
> to be comparatively but little noticed by the steam-boat passenger between Gainsborough and Hull. This is in many instances caused by the difficulty of gaining information in the midst of animated discussions on politics, farming, trading, scandal and religion . . . At the present time the town of Gainsborough possesses a most important branch of British commerce, the carrying trade. It has lately been ascertained that upwards of 2,000 tons of goods pass through it weekly, and the dispatch used by the various wharfingers in forwarding articles to their destination has been very materially accelerated by use of steam-vessels.[55]

CRINAN CANAL

NOTICE.

CHILDREN and Others are hereby Prohibited from Running along the Canal Banks after the Passenger Steamer; and Passengers are requested not to encourage them by throwing Money on to the bank.

Children are further warned not to throw Flowers into the Boat.

SALE OF MILK.

THE SALE OF MILK on the Property of the Canal Commissioners is only permitted on the understanding that no annoyance is caused to Passengers.

Any person who, by urging to purchase, or otherwise, inconveniences or annoys any Passenger will be prohibited from selling, and, if necessary, dealt with according to law.

Passengers are requested to report any such case to the Purser, and also to point out the delinquent to the nearest Lock-keeper or Canal Official.

L. JOHN GROVES,
SUPERINTENDENT.

Crinan Canal Office,
Ardrishaig, 30th June, 1887.

Fig. 10.

And in 1835 on the Shropshire Union Canal:

I made one voyage by this canal, from Chester towards Liverpool, by the packet-boat, which started from the canal basin at eleven o'clock in the morning. Notwithstanding that the bad navigation of the river Dee necessarily tends to increase the traffic on this canal, the indications of business, both in the office and on board the packet-boat, were fewer than might be imagined; that is to say, there were not more than twenty or thirty passengers, and the inland produce chiefly consisted of live fowls . . . The incidents on this short voyage were but common-place, though it behoved each passenger to exercise some degree of watchfulness, to prevent his brains being beaten out by the arches of the numerous bridges across the canal. These are so low as not to allow an individual to stand upright when passing underneath, and they are encountered at the rate of five or six in a mile. The boat was towed by three horses, of which a boy rode the hindmost, driving the other two before him without reins. The animals, free of controul [*sic*], were nevertheless, like men in the same predicament, not quite so independent as might be imagined, the towing-path being so straightened and narrow, they were unable to turn around . . .'[29]

Not all trips, of course, were found to be so boring, presumably it depended on how blasé one was towards the whole thing.

'Here we took possession of a spacious barge, with a house in it and a comfortable dining-room or drawing-room within the house, and a level roof on which we could sit at ease, or dance, if so inclined. These barges are common at Oxford – some very splendid ones being owned by the students of the different colleges, or by clubs. They are drawn by horses, like canal-boats; and a horse being attached to our own barge, he trotted off at a reasonable pace, and we slipped through the water behind him, with a gentle and pleasant motion which was like no motion at all . . . In this happy state of mind and body we gazed at Christ Church meadows, as we passed, and at the receding spires and towers of Oxford, and on a good deal of pleasant variety along the banks; young men rowing or fishing, troops of naked boys bathing, as if this were Arcadia, in the simplicity of the Golden Age: country houses, cottages, waterside inns, all with something fresh about them, as not being sprinkled with the dust of the highway . . . Meanwhile a table had been laid in the interior of our barge and spread with cold ham, cold fowl, cold pigeon-pie, cold beef, and other substantial cheer, such as the English love, and Yankees too – besides tarts and cakes and pears and plums – not forgetting, of course, a goodly provision of port, sherry, and champagne, and bitter ale, which is like mother's milk to an Englishman, and soon grows acceptable to his American cousin'.[56]

And at the other end of the country: 'In sailing down the river Irwell, by the elegant and commodious Pacquet-Boat, which starts from the New Hayley Bridge, Manchester, the passenger is highly gratified with many romantic and picturesque scenes . . .';[57] and even from Cobbett's 'Great Wen' it was stated that: '. . . the cheap trips into the country offered by its means during the summer months are beginning to be highly appreciated, by the people, who are pent in close lanes and alleys; and I have no doubt the shareholder's dividends would not be diminished by a more liberal attention to this want'.[58]

Although the packet-boat trip could well be efficient, comfortable and, if the weather was fine, interesting, it could also produce problems.

'The navigation of the Ouse and Humber, owing to shoals and shifting sands, is as bad as can well be, at all times . . . Notwithstanding such a state of things, and although it is impossible to make the voyage down unless with sufficient water, and at the top of a tide, the proprietors of the packet-boat start every morning unflinchingly at eight o'clock, being the time of arrival of the railway train – this in spite of wind, and tide, and in defiance of all rational objections. Punctually at eight o'clock the Leeds train arrived, with a numerous cargo; when all the passengers and luggage were put on board the steamer intended to carry us to Hull. Doubts were soon expressed by those partially acquainted with the river . . . but before we had been a couple of hours on the way, indications appeared sufficient to set speculation at rest . . . and two men, one on each side, each with a chequered pole in his hands, continually announced the soundings. We were tantalized for some time by hearing "six foot, five foot, five and a half foot, five foot" and so on, till at last came "four and a half foot" and then she stuck . . . the captain . . . received able support from all his passengers, who, about forty in number, condescendingly acted in concert under his directions, and shuffled across from one side to another so as to keep her going, and prevent her from lying quietly down on the mud . . . At the same time they hauled upon a rope, previously sent on shore, and made fast to a purchase, till the vessel was disengaged from her soft bed, and again afloat in a channel nearer the shore . . . We proceeded now about two miles farther, when the men with the chequered sounding poles were at work again for a few minutes, and then came an end of all uncertainty, for we touched the ground again, and in a few seconds were laid up in right earnest . . . here she remained for three hours, during which time an opportunity was afforded to those inclined to reflection to determine the cause why this packet boat might not, by starting some time later, have allowed the people to pass their time at Selby, instead of upon this mudbank. On asking eagerly for information on this point, it was hinted that the liquors on board were excellent; but this is mere hearsay . . . At all events a similar fate to the above related betided the passengers who arrived the two subsequent days at Hull; and an iron passage-boat recently placed on the station, notwithstanding her lighter draught of water, meets frequently with like disasters.'[29]

For a time thereafter the only real interest in waterway transport for passengers came from boat or skiff owners on the Thames and other rivers. Among the manifold duties of Mr Polly at the Potwell Inn were '. . . embarking, landing, and time-keeping for the hirers of two rowing boats and one Canadian canoe, bailing out the said vessels and concealing their leaks and defects from prospective hirers, persuading inexperienced hirers to start down-stream rather than up, repairing rowlocks and taking inventories of returning boats with a view to supplementary charges . . .'[59]

A few gentlemen had the time and means, or inclination, to either own their own craft or to hire one from a carrying company for holidays or outings. But only a very few boats were kept as pleasure craft, for paid holidays, other than the religious festivals, are a relatively new thing for the British working man. There were, however, occasional jubilees, school outings, church outings and trips, sometimes paid for by the squire. Some involved the use of five or six boats, together with their attendant boatmen and horses, taking the entire population of a village – the idiot came too – for a day upon the canal. Even if it were but a trip of ten miles, this might be further than some of the locals had ventured before. Odd notes appear in church magazines – 'Boating today, using two of Tonks boats – clean, food good, returned 7 o'clock' – and bills would sometimes appear, 'To hire of three barges for outing by the day. Three at six shillings = 18*s*.' An instance of this type of pleasure-going on canals appears in a church magazine dated June 1901.

> It came off as usual on Whit Monday in glorious weather. St. John's and St. Michael's Sunday Schools wended their way to the canal bank, where five barges were in readiness to take them to Clydach. This year an unusual number of visitors joined us. After a very pleasant journey up the canal, the field was reached soon after one o'clock. By this hour the buns provided were most acceptable. By 3 o'clock tea was ready, and each class, with its teacher as attendant, sat down on the grass, and, needless to say, enjoyed a good meal. After tea all took part in different games, some going in for cricket, quoits, races, and other sport. A 7 we began to think of starting for home, and again our journey was quite pleasant. By 9 o'clock we were at home all having enjoyed a most pleasant time . . .[60]

By the turn of the century the rivers and broads of Norfolk and Suffolk were already heavily used by pleasure craft, sufficiently so for a guide to be issued. Interestingly, there were a set of regulations, admittedly unofficial, which could well be contemplated today.

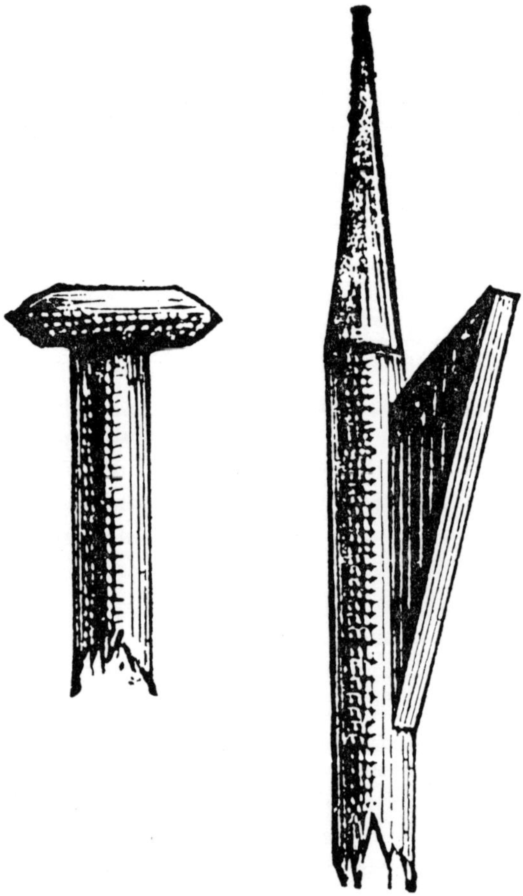

BUTT AND HEAD OF A QUANT.

Fig. 11.

Do not, in the neighbourhood of other yachts or houses, indulge in songs and revelry after eleven p.m., even at regatta times.

Bathe only before eight o'clock in the morning, if in sight of other vessels or moored in a frequented part of the river. Ladies are not expected to turn out before eight, but after that time they are entitled to be free from annoyance. Young men who lounge in a nude state on boats while ladies are passing (and I have known Norwich youths to do this) may be saluted with dust shot, or the end of a quant [boat shaft].

Adhere strictly to the rule of the road when boating . . . and when angling, moor out of the way of sailing craft . . .

Do not throw straw or paper overboard to float to leeward and become offensive; but burn, or take care to sink all rubbish.

Do not light fires, place stoves, or throw refuse on the banks in the path of others, whose yachts may be moored on the same bank.

Steam launches must not run at full speed past yachts moored to the bank, particularly when the occupants of the latter have things spread out for a meal.

Don't take guns on board unless you have leave to shoot on somebody's land.

Remember that sound travels a long way on the water and do not criticise the people you may encounter with too loud a voice.

Don't go on a friend's yacht with nailed shoes (the commodore of a Thames sailing club once came on board mine in cricket shoes armed with spikes). Don't knock the ashes out of your pipe into his boat, and don't catch small fish and litter his decks with them, leaving them for him to clean up after you.

Don't moor outside another yacht without the permission of its owner.

Ladies, please don't gather armfuls of flowers, berries, and grasses which, when faded, you leave in the boat or yacht for the unfortunate skipper to clear up. Don't play the piano in season and out of season (the reedbird's song is sweeter on the Broads); and don't turn out before eight o'clock in the morning when other yachts are near.[61]

Some of the above, which I suppose we should call recommendations rather than regulations, sound a little quaint. Not so many young ladies these days can play the piano let alone take it on a cruiser, but the treatment of yobs who bathe in the nude seems apposite.

In the years immediately preceding the First World War there was quite an upsurge of interest in canals both for commerce and for pleasure, some part of which may have been derived from the coming of the petrol engine. Mr P. Bonthron, an indefatigable canal- and river-cruising man, used a number of craft of which he was inordinately proud. The first was an open launch, measuring 27 ft by 6 ft 3 in. with a draught of 2 ft 9 in., powered by a 6 h.p. Daimler car engine, of which he said:

> '. . . our boat was well equipped against contingencies after our many experiences, and we had two coils and two accumulators, one set being fixed on each side; in addition to this, we carried a dry battery, good for many miles; then we had our tube ignition (which, however, was unsatisfactory), and, lastly, a stout towing mast – used also for the mast-head light – with a strong rope, so that we were in a measure prepared for any trouble that might arise. The last resort was, of course, the horseman . . . and in this again one enjoys a new and pleasant sensation after the vibrating engines, although, of course, this mode of progression is infra dig, to the motor boat tourist.'[62]

For his first reported trip he took a circular tour from the River Thames at Chertsey through the Thames & Severn, Stroudwater, Gloucester & Berkeley canals, the River Severn, Worcester & Birmingham, Stratford, Grand Union and Oxford canals and back to Chertsey, a matter of 380 miles and 291 locks.

> 'This circular tour may be pronounced a great success, and if one is so bent a more delightful holiday could hardly be taken. It dispels too, the erroneous impressions that some people have about canals and travelling thereon, as one passes through lovely country that it is the lot of very few to take advantage of. The great charm is that one is far from the beaten track, and human beings are few and far between. The

adjacent country generally is conspicuous by the absence of population for miles around. This we found also to be the case on our previous travels on the waterways in rural England. One is practically beyond the bounds of civilization, and yet still within hail of large towns. The ever-changing scenery keeps one alive all the way along, and the panoramic effect is very fine. In travelling thus one must put up with things as they are found, and be content at that; and all thoughts of "Metropoles" and "Cecils" must, of course, be left behind.'[62]

These latter, incidentally, were very 'U' hotels in London.

There were, of course, perils in a trip of this nature as there were in most of his trips, primarily the unreliability of the early motor boat and secondarily the finding of hotels at night, for although they may have 'gone canal cruising', the author and his companions, in the true Edwardian manner, declined to slum it at night. He recommends a number of hotels and states that for another round trip of 194 miles he used 24 gallons of petrol (Bowley's Motor Spirit), which contrasts well with outboard engine consumption and, perhaps, offset the hotel bills. Although the author occasionally has a snap at inefficiency, he says: 'I should like to say that wherever we went we were shown the greatest kindness from everyone we came in contact with, the lock-keepers included, all through our tour.'[62]

For the Basingstoke Canal Mr Bonthron used his skiff, thus emulating *Three Men in a Boat*, railing the skiff to his pick-up point, where he was pleased to note that he found it made ready for his immediate use. Other little side trips he took to canals which were no longer navigable were undertaken in a 14–16 h.p. open Darracq Tourer. He also possessed a rather fabulous houseboat *Clytie* which he kept on the Thames, complete with a maid and flowers in the window-boxes. His was the gentlemanly way of wandering around the canals.

In 1911 there appeared a book by E. Temple-Thurston entitled *The Flower of Gloster*. This gentleman hired a narrow boat, an ex-Thames & Severn Canal boat, together with the boatman and went wandering off along the Thames, the Oxford and Stratford canals, the Rivers Avon and Severn and back along the Thames & Severn Canal, engaging *en route* in philosophical discussions with the boatman, Eynsham Harry, and enjoying the canals in the gentlemanly way.

In 1928 Mr C. Auberton wrote *A Caravan Afloat*. Not so hard-pressed as Mr Bonthron to cover the maximum mileage in order to get to an hotel, nor quite so leisurely as the philosophical Temple-Thurston, the Aubertons were, perhaps, the first pleasure craft people who would be recognisable as such today. Since the end of the Second World War the number of pleasure craft has grown enormously, with no less than a 50 per cent increase between 1967 and 1972. Pleasure craft in this context may be taken to mean those using canals other than for commercial purposes. However, within this classification there are a number of varieties. Firstly we have the old-fashioned packet boat, often horse-drawn, which may operate day trips from a fixed base, usually at a reasonable charge. A development is the hotel craft which may either operate from a fixed base and undertake weekly trips, or it may make its way around the network, changing passengers at various staging points.

Continuing on the hire-boat theme one comes to people who live too far from a canal to own their craft, or find it uneconomical to do so, and instead hire boats by the week or the fortnight, often returning to the same company and the same stretch of water many years in succession. The late Arthur Rice of the Kingfisher Line at Hoo Mill Lock, the owner of one of the oldest hire-boat companies, told us he had customers who came back to him every year for 20 or more years, often adjusting their holidays in order to get the craft they wanted. He explained that he only advertised in the *Daily Telegraph* and *The Times* in order to ensure that he has a select clientèle who will look after his craft and for this reason he can fit them out just that little bit better. Other companies, aiming at a wider market, find all sorts of odd things happening to them. Given a wet week it is not uncommon for people to abandon a cruiser, often miles from the nearest safe moorings, catch a bus or train back to the boatyard and sneak their car out in the hours of darkness. This treatment is normal if any damage has been done to the cruiser but in any case many prefer to forfeit their nominal deposit rather than struggle with rain, mud and fratching children. On one occasion a hire-boat operator, hearing nothing through the grapevine, traced his craft to the Welsh Canal and found it there, stripped of everything saleable. Another boat was brought back to the yard at 2 a.m. with a hole in the side, the engine damaged from having been run with no cooling water, vomit all over the cabin and beer bottles tinkling musically in the bilges. Another fibreglass craft was taken out for a week, and on the Wednesday the hire-craft owner was asked to go out to the Birmingham Main Line as there was a hole in the hull. 'Not uncommon' he thought and picked up a fibreglass repair kit. On the main line at that time one was only permitted to use one of each pair of arches on the double bridges, as the other side was not dredged. The hirer had taken the wrong side and hit a piece of scaffolding pipe. He blithely explained to the owner that he had then reversed, saw what had happened and tried to go through again, inevitably making another hole. The owner was more resigned than angry.

Some lock-keepers live in mortal fear of hire-boat companies who have 70-ft converted ex-working boats. Generally they are well and sensibly crewed, sometimes however, the crew will be made up predominantly of school children with one or two adults (nominally) in charge. There can be little less horrifying to a lock-keeper plodding peacefully up the towpath than to see 13 tons of motor boat with the engine still going at full speed approach his lock gate without a steerer being visible. Called to attend to a boat once which was parked on the top gate of a lock on the Grand Union Canal, with both gate paddles and the handrail broken, I ascertained that the steerer was an 11-year-old boy and the box he was standing on had collapsed at the wrong moment. He protested that 'it ain't got any brakes like a car'. The adult in charge just shrugged his shoulders and said 'Send the bill to the . . . Comprehensive School', obviously thinking that it was only a matter of a few minutes' work to repair the lock.

The bad points connected with boat hirers naturally stick in the mind more than the good ones, but for every bad steerer there are probably 20 or 30 good ones who pass unnoticed. Some, indeed, are the epitome of courtesy and others when faced with their first lock are only too glad to be shown how to operate it properly and follow the instructions thereafter. Most lock-keepers are not unwilling to explain and a keg of beer can assist to smooth the passage!

Private owners fall into much the same patterns. There are those who own robust, attractive craft and handle them with most of the skills of a boat's captain, but there are also those who own an imitation working boat, garishly painted on the exterior with roses and castles but rarely entering a lock for fear that they might get their hands dirty. Shafts and boat lines are conspicuous by their absence but the polish on their brass and copper is dazzling to the human eye. These are the canal equivalent of the pre-war 'Chromium-plated Maidenhead Cruisers'.

In the smaller boat line those which tend to pass unnoticed are usually owned by the more impecunious of the canal fraternity. They tend to take things as they come but others are always in a hurry, and are loathed by all except their own kind. In order to obtain a summons for speeding on waterways it is, at present, necessary to have two independent witnesses and to time the craft over a pre-arranged mile. For a number of reasons it is to be hoped that one day the British Waterways Board will be able to afford to introduce radar detectors against the speedsters and prosecute them, much as the police do with such efficiency on the roads. The offenders rarely have licences and more often than not have their craft trailed to an obscure wharf. While a standard working boat requires a 20 h.p. diesel to move itself and a laden butty with 50 tons of coal on board, 'Noddy Boatmen' require from 40 to 90 h.p. outboards to propel their 16-ft lightweight plastic boats. They will be recognised foregathered at some boat marina bar by their clean hands, white shirts and small 'boat-captain's hat'. Their lack of licences, of course, reflects upon every other canal user, for the British Waterways Board's revenue, and hence expenditure on amenity canals, is governed by the amount they receive in licensing dues per annum.

Another of the Board's problems in this context is revealed in their news release dated 17 January 1972, which states:

> On behalf of the British Waterways Board, the British Transport Docks Board's Research Station at Southall are carrying out research on their canal and waterway network. The object

50 The Ashcroft family with one of their boats, Lancaster Canal, *c.* 1890

51 Wide boat, Lancaster Canal, *c.* 1900

52 Hereford & Gloucester Canal near Asperton, 1965

53 Old Sod House Lock, Market Weighton Canal, 1969

of the research is to pinpoint the causes and possible cures of the wash from pleasure craft. The study will concentrate on hull form and propulsion. The waterway network was, of course, constructed before the advent of powered craft and although the British Waterways Board's Byelaws restrict canal boats to 4 m.p.h., damage can be caused to the banks, even at that speed, by wash created by motorised boats. This adds to the Board's already large dredging task and of course increases the expense.

A sensibly steered boat, moving at the legal speed can cause little, if any, damage but our 'Noddy Boatman' is oblivious to any damage he may cause, looking with pleasure at the breaking wave following him. This all too quickly undermines the banks, leading to leaks and silting up of the fairway.

If canals were widened to a commercial standard, pleasure craft would find their lives a little easier, for having the greater draught of water they would move quicker, locks would have to be filled mechanically and propellers would not become so easily fouled with extraneous matter. The contrast in the boat's movement on the Staffordshire & Worcestershire Canal and the River Severn is amazing to the unsuspecting.

At the commencement of this chapter the possibility of filling in canals was mentioned. This is, not unnaturally, the dream of many civil servants, but alas for them, an Arcadia impossible to achieve. If a canal is closed, where do you divert the water, whether from field, housing estate or river? Furthermore, what of the factories, power stations and water boards who rely on a supply? And it is not generally realised just how much of our domestic drinking water is taken from canals – one does hasten to add *before* purification!

In 1965 the British Waterways Board issued a study of certain waterways from which certain salient facts and figures can be garnered.[63] Incontrovertibly, the costs given must now be increased; doubling them would not be an unrealistic estimate. In their report the Board upset some, and pleased other groups of people with their statement: 'In the development of industrial processes and methods (including transport media) it is usually and rightly accepted that natural selection works. In other words, it is assumed that if a process disappears or virtually disappears, then that is because it suffered from certain insufficiencies which prevented it from competing successfully with younger and stronger breeds. The history of narrow boat carrying is one of long-continuing decline.' The report went on to give the reasons – admittedly logical – for the encouragement of pleasure craft. To totally eliminate a ten-mile length of a rural narrow canal was estimated to cost £6,000 a mile, the greatest single element in which (one-third) was bulldozing the banks and importing fill. Culverts, alternative water supplies and pipe discharges accounted for another £1,700 of the total.

By contrast a rural wide canal would take £9,000 a mile, bulldozing now taking £4,000, while culverts, alternative water supplies and pipe discharges would rise to £2,000. In cities and towns, the canal in question would, they say, 'require case by case treatment'. By far the most frightening factor is that even when the canal is eliminated a considerable expenditure would be incurred for a long period, at least ten to twelve years. Given below is some indication of the relevant figures.

Canal or Navigation	*Mileages (1965)*	*Conversion to a non-navigable water channel; deficit per annum*	*To eliminate* Interim *(over 10–12 years) Deficit per annum*	*Final*
		£	£	£
Ashby de la Zouch	22¾	7,500	9,900	7,900
Birmingham & Fazeley	22	9,100	9,700	6,100
Bridgwater & Taunton (Un-navigable since 1940)	14½	3,100	10,400	5,300
Coventry	27	4,700	15,100	7,300
Erewash	12	600	7,500	6,000
Grand Union Lock 7 (Buckby to Birmingham)	47	31,300	37,200	25,000
Kennet & Avon Bath–Reading	43¼	26,300	29,000	23,300
Leeds & Liverpool Wigan–Leeds	97	59,400	71,700	52,300
Stratford-on-Avon (North)	12½	3,600	5,700	4,200

This report at least provided a good argument against closure and then showed that the deficit to the Exchequer even if all waterways were converted to water channels or eliminated – whichever was the cheaper – would still stand (in 1965!) at £600,000 a year. The difference between this undesirable state and keeping such canals as are usable for pleasure craft was then a pitiful £350,000 a year.

These then are the treatments that can be meted out to waterways in future years. It is my own opinion, and that of most of my colleagues, that canals must not be closed but enlarged or superseded by modern navigations. We cannot, nay, must not lose this heritage of ours; for 200 years canals have moulded and assisted Britain towards her greatness; it is not fanciful to visualise that both country and canal will, if left to do so, decline together.

Appendix 1

STOPPAGE AT LOCK 25, STOKE PRIOR, WORCESTER & BIRMINGHAM CANAL

PERIOD OF STOPPAGE:	11th–22nd October 1971
REASON FOR STOPPAGE:	Replacement of two bottom end gates, together with paddles, etc.
CRAFT UTILISED:	Motorised flat No. B67. Butty boat *Leo* (ex-Grand Union Canal Carrying Company, No. 300).
MECHANICAL EQUIPMENT AVAILABLE:	One hired 3 in. pump. One hired digger/excavator. One BWB-owned gantry.
PERSONNEL:	Two carpenters, one carpenters' mate, one bricklayer, one bricklayer's mate, three labourers (one of whom was also a driver), one foreman (part-time).

Prior to the actual stoppage a certain amount of work was involved, taking place mainly in the preceding week. On Monday and Tuesday the craft *Leo* and B67 were loaded at Tardebigge Yard and a baulk for use as a clap sill and part-assembled gates were brought from Bradley Workshops (Birmingham Canal Navigations) together with the gantry, by road. A number of men were employed on the unloading and subsequent reloading as they were available, while the carpenters prepared the new paddles, assembled new ironwork and cut wedges, packing pieces, etc. On Wednesday and Thursday the craft moved from Tardebigge to Lock 25, with seven men, three to handle the motor-boat and four bow-hauling (pulling by hand) the butty-boat through the 33 locks *en route*. On Friday two of the labourers proceeded to clean up the banks, hedges, etc., to allow free movement of men and materials up to and around the stoppage site. The bricklayers, carpenters and their mates moved the *Leo* into Lock 25, assembled the gantry and pulled it into position. The gantry ran on wooden baulks; wedges were laid down fore and aft of the gantry wheels to prevent unwanted movement.

Monday 11 October. Prior to moving off from Whitford Bridge, one set of stop planks, used to retain water, was dropped off, together with a small amount of ashes for 'racking-up' purposes. At Lock 26 a further set of stop planks was dropped off and the craft proceeded to Lock 25, where it discharged the stop plank for use at the top of the lock; the boat was completely stripped of all the equipment lying in the hold. The clap sill, a piece of elm measuring 12 ft × 12 ft × 7 ft long, weighing 3 cwt, was lifted out manually – only the gates were left in the craft. In the meantime B67 had dumped a set of stop planks at Hammonds Wharf and had taken on board a mechanical pump. A rough inventory of the material now on site would show, apart from miscellaneous packing, etc., two full-length ladders for access to lock chamber, three double walking-planks for crossing over the lock when it is dewatered, iron posts and their attendant lines for placing around the lock, the clap sill, gantry and six baulks per side of the lock, two boat shafts, a dismantled anvil-cum-vice, normally carried by the carpenters on *Leo*, tool boxes, 'horses', sawing-bench and trestles, two new paddles and their attendant iron-work, ashes, clay, wedges, iron pots for fitting to the new gates, etc.

At eleven o'clock a number of men were sent out on water control. Two pounds, those between Locks 28 and 27 and 27 and 26 were lowered between 18 in. and 24 in. to provide reservoirs to absorb water spilling over the weirs from higher up the flight. Despite the issuing of Stoppage Notices it was still obligatory on the Board to warn persons who had craft near any locks liable to be affected that there was a danger of grounding. The foreman and one labourer were engaged in cutting teeth from the bucket of the hired digger in order that it might lift the slurry from the bottom of Lock Chamber 27. The machine had previously been brought across the fields, uprooting various trees, bushes, etc. *en route* so that its movement would not be obstructed on the lock-side. The trees and bushes then served as a dam to stop the slurry from washing down into the farmer's field.

One labourer was engaged at Lock 25 unbolting the old iron-work from the bottom gates in order to lighten them, while the carpenters' gang sawed the balance poles off the old gates and put the lifting chains ready. The balance poles had to be sawn off to allow the gates to lie lengthwise within the *Leo*. Walking-planks were placed across the lock, together with walking-platforms, to give safe access; handrails were not fitted.

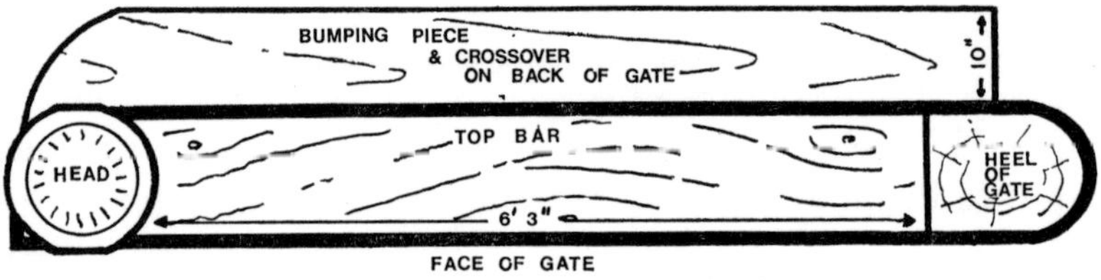

Fig. 12.

Due to the very poor condition of the outside bottom gate, the lifting chains were not easy to place in a secure position, but were eventually located underneath the walking-board. By one o'clock the gate was manoeuvred into the bottom of *Leo*, the heel being dropped in ever so gently and by gradually lowering as the boat was moved out of the lock the gate slid into its appropriate position.

During the afternoon two men manhandled the pump uphill from Lock 24 to Lock 25. By Monday evening both old gates were out, the chamber of Lock 26 cleansed of slurry, stop planks were in place at Whitford Bridge and Hammonds Wharf and the new balance poles taken out of the *Leo* ready for use at Lock 25.

Tuesday 12 October. Three labourers loaded a 2-in. pump for dewatering the lock plus safety helmets and other bits and pieces on to the flat at Hammonds Wharf, whilst the other men lifted both new gates out of the *Leo* and put them roughly in position.

By mid-day both balance beams had been moved and were

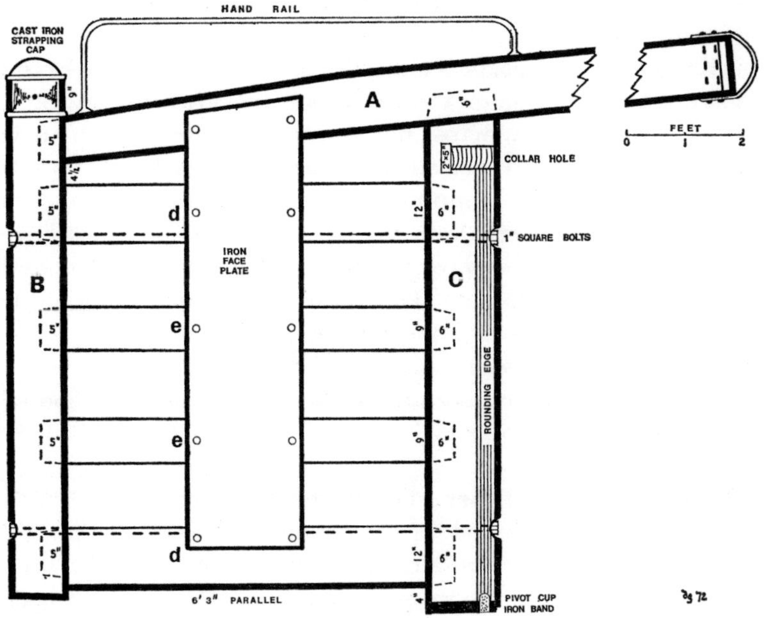

Key:

A. Balance beam. 20 ft overall and measuring 12 in. × 10 in. is sawn 6 in. out of straight the 10 in. way to throw the overhanging part of the beam away from the lock-side alignment when the gate is open. 'In the days when we did our own sawing we would pick a log that had a natural bend in it to suit the required crank.'

B. Headpost. 9 ft overall, measures 12 in. × 10 in.

C. Heelpost. 9 ft overall, measures 12 in. × 10 in. The tenons into the headpost are 5 in. deep, those into the heelpost 6 in. All tapered.

D. Cross-rails, 12 in. × 10 in.

E. Cross-rail 9 in. × 10 in.

2 in. face-planking throughout.

Fig. 13. A 'top-end' gate to suit any lock from No. 3 to No. 58 inclusive, Worcester & Birmingham Canal. Based on a drawing by G. D. Bate, B.E.M., Foreman Carpenter.

in position on the gates, the procedure being that the tenon (pronounced tenant) was located in the mitre-post and wiggled into the hole, the mortice was lowered and any final adjustments were carried out with a sledge-hammer. All the labour force was by now assembled at Lock 25, but regrettably only five partook of the work involved in moving gates and balance beams. The foreman arrived mid-morning and all the material that had been carefully unloaded from the *Leo* had, due to inefficient planning, to be moved to new locations. The digger, after picking up the old sleepers and piles used to cross the lock, was escorted by the foreman and two labourers down the towpath, where it placed piles across the lock and assembled in its new position proceeded to excavate the chamber of Lock 25.

By two o'clock the pump had been positioned ready to pump out the immediate area. Stop planks were now in position above Lock 25 and the areas between Locks 27 and 26, 26 and 25 and 25 and 24 were dewatered. The *Leo* was left in the centre of the channel below Lock 25, the flat B69 being on the opposite (offside) point between Lock 24 and 23, the water-level of which was maintained because of the number of craft moored at the pleasure boat base.

Wednesday 13 October. The digger the previous afternoon had cleaned out only half of the chamber at Lock 25 and did not finish the job until 10 a.m. Most of the men arrived a little late, but the vacuum pump was already running, having been set going by a labourer who lived at Hammonds Wharf. The first job was to clear out the chamber and apron at the bottom end, prior to fitting the gates. Most of the slurry was shovelled into the bucket of the excavator and lifted out that way; surplus water, etc., was scooped out with an instrument the name of which is pronounced 'scope'. In the meantime a pump had been set going to pump out any water which worked through the stop planks at the top end of the lock. This, and shepherding the digger when it moved off to Lock 24, was to be the sole work of two men for the day. By ten o'clock the lock chamber and general area were clear and the gates moved over by gantry to allow access to the rubble, etc., in the paddle holes. Two windlasses were found.

After the morning break work started in preparation for the demolition of the old clap sills and the pivot boxes, which proved to be rotten where they were set into the main sill. The weather was bitterly cold at this point and work proceeded slowly. At eleven o'clock a foreman arrived and detached the labourer/driver for other duties. The digger made its way down the towpath to Lock 24, crossed over and commenced work clearing the lock chamber. The other workmen continued as planned, one carpenter cutting the clap sills to size and shape.

At twelve o'clock the rain which had been coming down quietly fell hard enough to drive the men under cover and, with one small exception, that was the end of the day's work. The exception was when the foreman and labourer/driver loaded a further set of stop planks and placed them in position above Whitford Bridge at Lock 26. The rain continued heavily and it was anticipated that there might be difficulties with excessive water.

Thursday 14 October. The weather continued cold and overcast, but all labour was on site with the exception of the labourer/driver who was posted elsewhere. Work continued with fitting the pins and boxes for the gates and the gates were then relifted with the gantry and lowered into place; the boxes, incidentally, are cut from elm and their replacement was necessary as the previous cups had worked loose and become oval – they were installed at the same time as the gates, 1917 and 1925. The new boxes were countersunk into the old main sill, caulked in with oakum and hot pitch and the cups inserted and treated in like manner (the old ones were discarded). Having finally got the gates into position they were prepared for the mitre cut. The majority of labour was involved on this. Using pumps at the head and tail of the lock, one labourer kept it clear of water. The bricklayer and his mate were engaged in minor repairs to the lock walls and various of the available men were detached at odd times to check the condition of the stop-planks, level of water in the pounds, etc. The hired excavator, together with one labourer, continued to clear out the lock chamber at No. 26; the operator afterwards turned his attention to the intervening lock pound which he was unable to reach with any success, due to the relatively short reach of the crane.

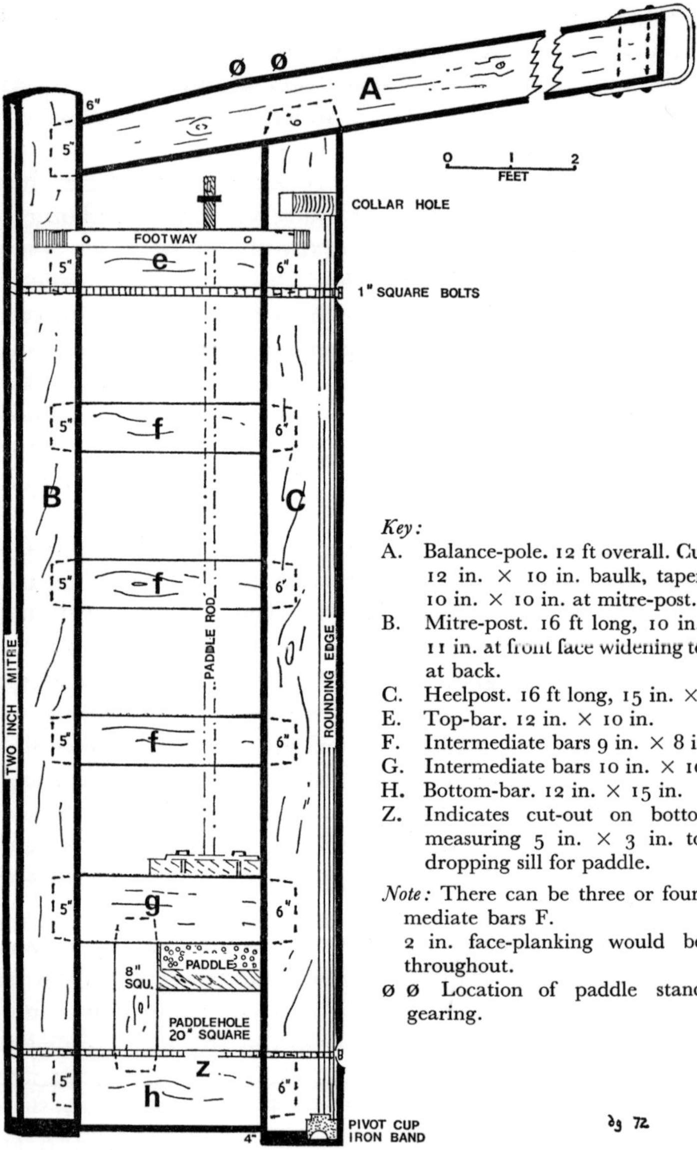

Key:

A. Balance-pole. 12 ft overall. Cut from 12 in. × 10 in. baulk, tapering to 10 in. × 10 in. at mitre-post.

B. Mitre-post. 16 ft long, 10 in. deep, 11 in. at front face widening to 13 in. at back.

C. Heelpost. 16 ft long, 15 in. × 10 in.

E. Top-bar. 12 in. × 10 in.

F. Intermediate bars 9 in. × 8 in.

G. Intermediate bars 10 in. × 10 in.

H. Bottom-bar. 12 in. × 15 in.

Z. Indicates cut-out on bottom-bar, measuring 5 in. × 3 in. to form dropping sill for paddle.

Note: There can be three or four intermediate bars F.
2 in. face-planking would be used throughout.
Ø Ø Location of paddle stand and gearing.

Fig. 14. One of the 'bottom-end' mitre lock-gates, Worcester & Birmingham Canal. Based on a drawing by G. D. Bate.

Friday 15 October. The majority of the labour, with the exception of the pump-man and the labourer/driver, were involved in cutting the gate mitres. When the gates were first dropped into place the minimum amount possible to make a watertight seal was cut away from the mitre-posts. The gates arrive in a rough sawn condition and having cut one post to the right size with a two-man manual cross-cut saw, the second gate had its high points indicated by putting raddle (venetian red) on the face. The gates were then closed and the high points, indicated by the raddle being transferred, were adzed away. In the case of Lock 25, however, the cross-cut saw, operated by two carpenters, was used for both cuts, mainly due to the irregularity of the face. As the cross-cut saw is 6 ft long the gate was held in the open position and ladders placed against each face, firmly secured at the top and bottom, and each carpenter ascended his own side of the gate and was retained by a labourer holding him there. The adage 'one hand for the job and one for yourself' cannot, in these circumstances, apply. Having completed this, each carpenter cut his own half of the clap sill to size. Where the halves abutted to the quoin a notch was cut out, measuring approximately 4 in. × 4 in. × 8 in., which at a later stage had an insert made to fit the round nose of the quoin. These throw-away pieces should, in theory, if the joint is to be kept water-tight, be replaced every five years. Only one lock of the 58 on the Tardebigge flight had a sharp-nosed quoin and it was often cursed by boatmen as it was easy to catch. The digger-driver completed the cleaning out of Lock 24 but inadvertently lifted away the mud plank kept at the tail of the lock to prevent mud, bottles, etc., coming between the gates and the clap sill when they are not in use. The balance of the afternoon was spent in moving everything possible from the site due to anticipated vandalism over the weekend.

Monday 18 October. Several men were late and it was about nine o'clock before work could start. Even then the local labourer had not had any morning overtime authorised and had not, therefore, activated the pumps, with the result that until about 9.45 a.m. the site was not ready for use, although some tools were brought out. While pumping-out operations began the labourer/driver and another labourer fetched a good paddle from the yard in lieu of one that was found to be damaged and unloaded piles.

In the meantime the carpenters had recovered a clap sill that local children had dropped into the bottom of the lock and it was also found that 35 out of 100 bricks had gone missing. Incredibly, little else was damaged. The carpenters then lowered both clap sills into position, fitting them to one another, while a new paddle, a bag of cement and some bricks were brought up, plus a ladder for the use of the bricklayer and his mate who had to fill a hole in the lock wall, together with an extensive hole in the weir at the head of Lock 24. Another labourer spent most of the day with a hammer and chisel cutting away a rotten curb stone at the head of Lock 25.

The clap sills did not fit and it was not until about three o'clock that they could be persuaded to do so. The carpenters and foreman then took it in turns to bore four $1\frac{1}{8}$ in. holes in each sill, prior to inserting the gagg pins, a job which was completed the same day. At 3.30 p.m. the threatened rain dropped but work continued until 4.30 p.m.

A party of schoolchildren with their teachers visited the site despite the rain, a rather dangerous proceeding as the children persistently ducked under the safety ropes.

Tuesday 19 October. The pump man having been paid his hour's overtime the site was ready for immediate occupation. Most of the men were involved in 'barring' the gates, which was done with an implement like a heavy crowbar with a hooked or oblique end; this is pressed into the brick-work at the bottom of the lock chamber and the gates are then levered forward. Next the 'bars' are set in position by means of two specially cut pieces of timber having a 'v' notch at one extremity. The idea was that the bars would hold the gates and the gates hold the clap sills in place. Unfortunately, the gagg pins were knocked in with sledge-hammers which moved them apart about $\frac{3}{16}$ in. and it was necessary for one carpenter to cut pieces of elm to this size while the other inserted oakum between the clap and main sills to make a watertight seal.

To pass the time one labourer was engaged in trundling

barrowloads of ashes from a dump at Whitford Bridge to the various locks, as insufficient had been left previously.

At approximately eleven o'clock it began to rain but it was not until twelve o'clock, when all the men on the site were soaked, that the carpenter, who was acting-foreman, called a halt to the work. Everyone retreated to the cabin of the *Leo* and, led by the carpenter, groused about the rain until 2.30 p.m., when the men proceeded to make up a mortar mixture and cut out an extension to the area left in the clap sills to take the throw-away portion. There was no apparent reason why this was done on site apart from a slight miscalculation. The bricklayer then used the cement mix, which was lowered to him a bucketful at a time from the cement mixer standing above the tail of the lock, to make good the brick chamber and as a grouting or back-fill to the clap sills, incidentally screening the misjoin on the clap sills. Having put a very good finish on the surface of this, a number of attempts were made by one of the labourers to cut his initials in the surface of the cement, but due to the presence of both gravel and cement fondu (quick-hardening) he only succeeded in making a mess. Two half-penny pieces were also inserted, rather pointlessly, as they were dated 1955 and 1960 respectively, and the bricklayer neatly cut his initials in (A.W.). One carpenter had cut his quoin fit (throw-away piece) and drilled and nailed this into place in the meantime, and proceeded, while the other men were clearing up, to assist the other carpenter. The quoin fits were made and put in place towards evening.

Wednesday 20 October. The weather was pleasantly warm and tempers became so when it was found that although the gates were a good fit to the clap sills and to one another the rubber inserts which make a watertight seal for the length of the gate against the quoin were not touching it. It was, therefore, necessary to lift the gates with the gantry (the inside gate also being rather less than satisfactory), dig out the pivots and relocate them approximately ½ in. to one side, which involved cutting away the old elm around the pivots, moving them and repacking. Minor complications were found in so far as the collars had to be slightly readjusted and the cotter pins holding

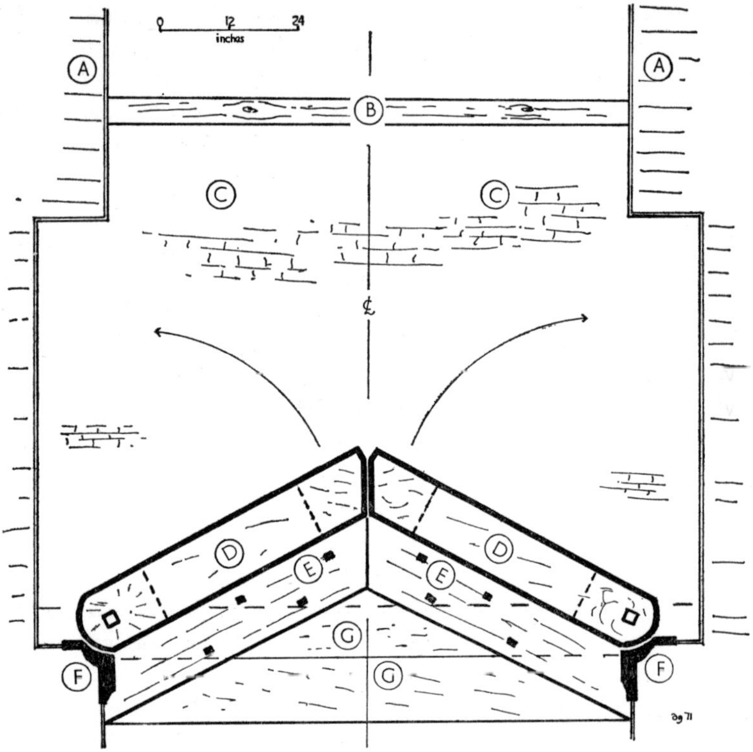

Key:
A. Lock walls, either brick or masonry.
B. Mud or rubbish plank.
C. Gate chamber, bottom usually brick.
D. Bottom gates, usually oak.
E. Clap sills, elm. Eight, ten or twelve 'gagg' pins (rag bolts) inset.
F. Iron quoins.
G. Main sills, elm.

Fig. 15. 'Bottom end' of a narrow lock, Worcester & Birmingham Canal. Based on a drawing by G. D. Bate.

the collars in place proved difficult to extract. Therefore the two carpenters and most other men at various times were engaged solely in this work, together with the fitting of one gate paddle and general fitting round of the gates. The resettlement of the gates meant, of course, that the mitre had to be recut once again. While the carpenters' gang were busy at this the bricklayer and a couple of other men were engaged in repairing the weir at Lock 26 and cutting out, ready to rework, some faulty brickwork at the tail of Lock 25. Two labourers had to take one paddle to the yard, because although it had been apparently cut to the correct size it did not fit. Four was found to be the absolute minimum number of men necessary to operate the power-saw in the yard.

On returning to Hammonds Wharf and having wheeled the paddle up to Lock 25, the two labourers were then dispatched again to the yard to collect 3 cwt of cement and two bags of building sand and with instructions to fill the crew bus to the roof with loose sand and gravel. With a payload of one ton this was not possible, but 16 cwt was taken to Hammonds Wharf, where it was shovelled on board B67 which had been brought down by two labourers for this purpose.

In the afternoon the lock hangers were assembled. These comprise two baulks which are placed across the lock and have hanging from them by chains two further baulks which drop inside the lock and are wedged into position with planks placed across to make a good working platform. At this point it was found that one of the carpenters had gone partially deaf, and an accident was only averted by someone's quick thinking. The lock hangers were for the benefit of the bricklayer; unfortunately in order for him to get at the brickwork it was again necessary for him to lift the outside gate.

The rigid pump hoses were taken down from Lock 25 to Lock 24 by four labourers, ready for pumping out that lock chamber, but due to a change of plan they were then carried back up to Lock 25. Of the material brought in the van none was used this day. Fairly heavy rain had caused the two lock pounds above Lock 25 to rise dangerously and water was run off for about an hour by removing the top stop plank and lifting the second to give a gap of about 2 in., the intention being to drop the level of both pounds by approximately 12 in. (this water would, of course, have to be pumped out by the pump-man on Thursday morning).

Thursday 21 October. Weather overcast, all labour present although the driver made three trips to Tardebigge for materials. The first load comprised 1 cwt of cement, 7 cwt sand and gravel,

3 cwt builders' sand and bits and pieces; the second, another bag of cement and the third trip was for an old scrap door to be used as shuttering. In the meantime the carpenters, together with two labourers, were putting into place the new paddles for the gates, with paddle-rods, paddle-rod cramps, etc. This kept them busy for most of the morning, although at one stage all available labour was engaged in manhandling the 3-in. vacuum pump down the towpath to Lock 26, where it was set going after stop planks had been placed in the bottom end grooves. Due to the rather peculiar shape of these grooves some difficulty was experienced in getting the planks in place, the operation taking two hours in the afternoon and involving six men standing on them to drop them into place. Lock hangers were also put in. After the pump was moved the bricklayer and his mate completed the brickwork required at the weir at the head of Lock 26. After the pump had lowered the water from the bottom end chamber it was seen that there was a foot of mud, in which the carpenters flatly refused to work; however, one labourer occupied himself for an hour going around the bottom with a shovel seeing what he could find, ending up with one paddle key. Due to the conditions the site was vacated at approximately four o'clock.

Friday 22 October. The weather was warm and sunny. The bricklayer had been detached for duty as relief yacht-basin attendant at Diglis (normally the bricklayer's mate would do this). The van driver was detached in the afternoon to dry-dock a boat at Tardebigge yard. In the morning, however, the pumps were running on arrival and the carpenters were engaged, while waiting for the water to drop sufficiently, in various odd bits of repair-work around Locks 25 and 26.

At Lock 25 one of the curb stones previously cut away was to be replaced by concrete. This was done by one of the labourers. The shuttering which had been brought out so urgently the day before was strutted into place and the hole filled. When the foreman arrived at mid-day he expected to see this dry, but the labourer who had done the job had not added enough cement fondu to the mix, fearing it would go hard before he could use it.

54 George Bate, B.E.M., retired Foreman Carpenter, Worcester & Birmingham Canal

55 Bill Harbach, Worcester & Birmingham Canal

56 Doug Hill, Grand Union Cana[l]

57 Steering the *Clematis* to Broad Street Depot, Wolverhampton, 1964

58 Jack James, master boatman steering *Serpens* near Cumberlan[d] Turn, Regent's Canal, 1969

At 10.15 a.m. two labourers had begun shovelling the mud from the bottom and chamber of Lock 26, but all work stopped for 20 minutes at eleven o'clock to allow a labourer to extricate an eel he found, which he proposed to take home and eat – after skinning it alive in the traditional canal manner. Notwithstanding this interruption the paddles were eventually 'fitted around' by 12.30. The process for 'fitting around' a paddle is that the nuts holding the paddle-rod to the paddle are undone, the paddle rod lifted out of the way slightly, the one paddle slide removed by knocking out the cotter pins and pulling it away. The paddle is then removed from the frame and placed on a 'horse box', generally cleaned up and inspected, the areas of greatest wear being fairly apparent. The paddle frame is then wiped clean and raddled, the paddle – a weight of about 56 lb. – being placed back and moved against the raddle. In theory, if it is a perfect fit, an area of 3 in. around the face of the paddle will be red. In the case of these two paddles the outside one was worn slightly but only took about ten minutes adzing and three placings against the paddle frame to get a reasonably good fit, but the inside paddle was very badly worn in the middle and took about 35 minutes. It is understood that chippies' rheumatism, common to most canal carpenters, was probably caused by holding paddles in place with the knee – the paddles are, of course, soaking wet, muddy and covered in lichenous growth. This job required two chippies, two labourers, the pump-man to watch from above and the eel-eater likewise.

During this time one other labourer was occupied in scooping the water out of the *Leo* prior to refloating, while the foreman was lifting the planks at Hammonds Bridge.

The afternoon was spent loading both craft with all loose material, the flat B67 towing *Leo* down to Stoke Prior Wharf where both craft were tied up for the night. The men did not leave the site until just after five. Some delay occurred due to the presence of craft moored against the towpath, but after removal of the last set of planks at the head of Lock 27 and removal of the chains which prevented egress of craft at this lock the stoppage was officially declared over.

Unfortunately, although this was the official end of the stoppage, further work was necessary on the site, for apart from dismantling the gantry and reloading and moving the craft it was found that the rubber inserts on the gates still did not abut correctly to the quoin and had to be removed and reset, which took the carpenters a few days. At Lock 26 a severe leak was found in the field behind the towpath. It came from the lock chamber, where it was apparent that some brick and stone courses had shifted, either from the digger moving across the lock or from its bucket having caught on the brickwork. The remedial works, apart from involving a large quantity of material, took a couple of workmen three weeks to complete.

No attempt can be made at costing this job, although the labour cannot, even with the then labourers' wages of £15·37½ per week, have been less than £500.

Appendix 2

Canals can run with, or against, the lie of the land, and in no way can this be shown more clearly than by their relation to the contours. In these maps the contours, locks, etc., have been simplified for clarity. All are based on Ordnance Survey Maps with the sanction of the Controller of Her Majesty's Stationery Office (Crown copyright reserved).

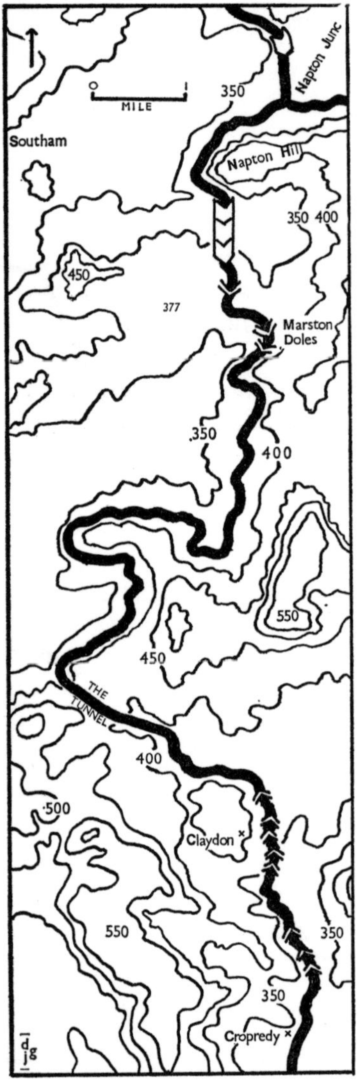

Fig. 16. Eighteenth century: The Oxford Canal.

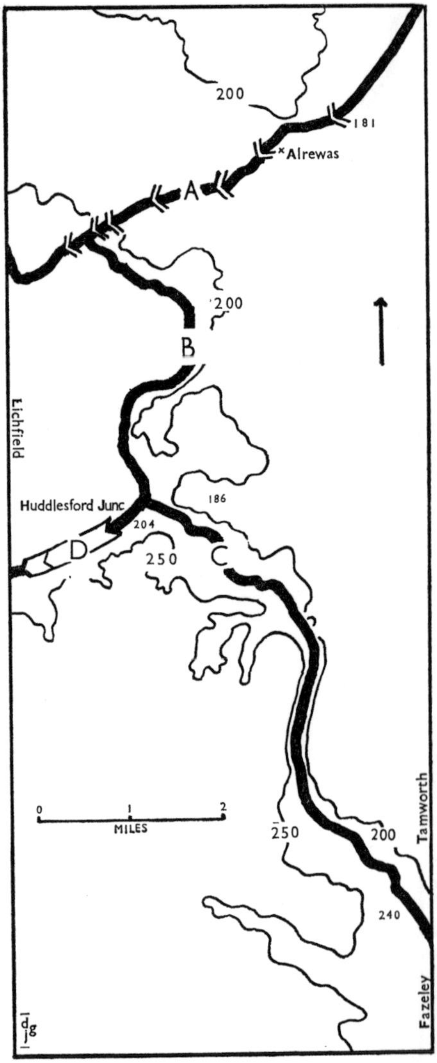

Fig. 17. Eighteenth century: A – Trent & Mersey Canal. B – Coventry Canal. C – Birmingham & Fazeley Canal. D – Wyrley & Essington Canal.

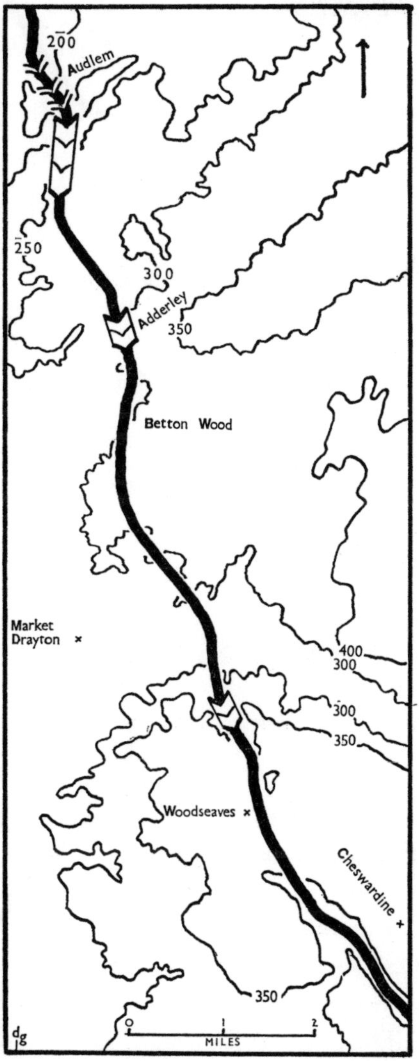

Fig. 18. Nineteenth century: Birmingham & Liverpool Junction Canal (Shropshire Union Main Line).

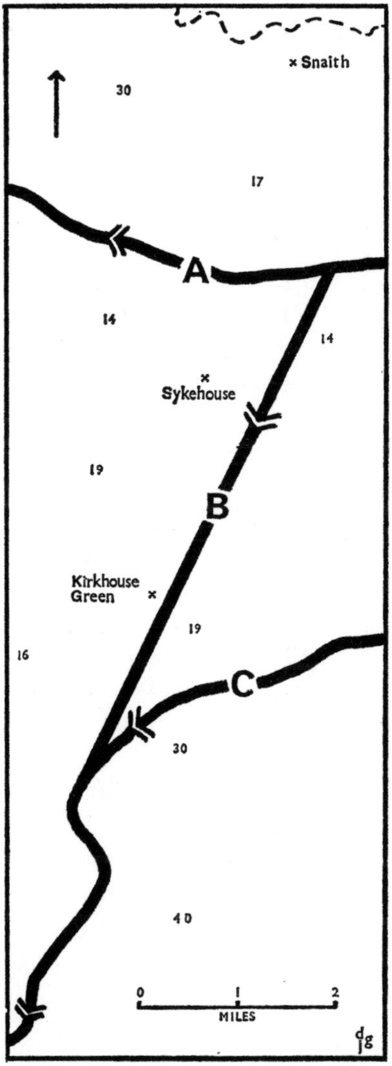

Fig. 19. Twentieth century: A – Aire & Calder Canal. B – Aire & Calder and Sheffield & South Yorkshire New Junction Canal. C – Sheffield & South Yorkshire Navigation.

References

1 HULBERT, C., *A History & Description of the County of Salop*, 1837.

2 RICKMAN (Ed.), *Life of Telford*, 1838.

3 MITCHELL, B. R., and DEANE, P., *British Historical Statistics*, 1962.

4 PRIESTLEY, J., *Navigable Rivers and Canals*, 1831.

5 YOUNG, A., *Six Months' Tour through the North of England*, vol. III, 1771.

6 SOPWITH, T., *An Account of the Mining Districts of Alston Moor*, 1833.

7 CRESY, E., *History of Engineering*, 1833.

8 British Waterways Board, Southern Region Notice No. 13(71) dated 29 December 1971.

9 M'CULLOCH, J. R., *A Dictionary of Commerce and Commercial Navigation*, 1852.

10 Southampton Record Office D/PM 6/3/11–13, D/PM 6/2/1 87–92, D/PM 6/11/6 and 7 D/PM 6/8/16.

11 Act of Parliament for the Stroudwater Canal, 1776.

12 *Essex County Chronicle*, 10 November 1818.

13 Berkshire Record Office, D/07 Bundle 1.

14 SHAW, S., *A Journey to the West of England in 1788*, 1789.

15 MACGILL, P., *Children of the Dead End*, 1914.

16 RAVEN, M., and RAVEN, J., (Ed.), *Folklore & Songs of the Black Country*, 1965.

17 TOWNSEND, T., (Ed.), *News of A Country Town*, 1914.

18 *Aris's Birmingham Gazette*, 6 November 1769.

19 *Aris's Birmingham Gazette*, 27 October 1769.

20 *Aris's Birmingham Gazette*, 23 July 1770.

21 WHITE, F., *History, Gazetteer & Directory of Cheshire*, 1860.

22 WARD, J., *The Borough of Stoke-upon-Trent*, 1843.

23 SMITH, G., *Our Canal Population*, 1875.

24 METEYARD, J., *Life of Josiah Wedgwood*, 1866.

25 Oxford Canal Records, 4/110/16.

26 TUFNELL, G., *Report on the Mines in Lancashire*, 1833.

27 ROBERTSON, H. R., *Life on the Upper Thames*, 1875.

28 CHADWICK, E., *Report on the Sanitary Condition of the Labouring Population*, 1842.

29 HEAD, SIR G., *The Manufacturing Districts of England*, 1835.

30 REES, A., *The Cyclopaedia*, vol. VI, 1819.

31 GLADWIN, D. D. and WHITE, J. M., *English Canals* – Part III *Boats & Boatmen*, 1969.

32 PYE, C., *A Description of Modern Birmingham*, *c.*1820.

33 *Aris's Birmingham Gazette*, 24 July 1769.

34 BENTLEY, W., *History, Guide & Directory of the Borough of Worcester*, 1842.

35 *Birmingham Post*, 21 December 1960.

36 OVERTON, G., *A Description of the Faults or Dykes of the Mineral Basin of South Wales*, 1825.

37 'A Description of a Coal Waggon Way &c', *Commercial & Agricultural Magazine*, October 1800.

38 SANDARS, J., *A Letter on the Subject of the Projected Rail Road between Liverpool and Manchester*, 1824.

39 *Lincoln, Rutland & Stamford Mercury*, 22 July 1854.

40 *Lincolnshire Chronicle*, 12 October 1855.

41 *Lincolnshire Chronicle*, 21 December 1855.

42 *Quarterly Papers*, vol. I, part 2, 1844.

43 Report by Committee on Criminal & Destitute Juveniles, 1852.

44 *Aris's Birmingham Gazette*, 2 November 1769.

45 TURNER, S., *A General View of the Agriculture in Gloucestershire, 1794.*

46 *Aris's Birmingham Gazette*, 19 November 1770.

47 Anon., *A Cursory View of the Advantages of an Intended Canal from Chesterfield to Gainsborough*, 1769.

48 *Leicester Journal*, 29 March 1805.

49 YOUNG, A., *A Six Weeks' Tour through the Southern Counties of England*, 1796.

50 WHITE, F., *History, Gazetteer & Directory of Nottinghamshire*, 1832.

51 EMERSON, R., *On English Lagoons*, 1893.

52 JEANS, J. S., *Waterways and Water Transport*, 1890.

53 Canal Control Committee (BOT) Handbook on Canals, 1918.

54 The *Marple Ratepayer*, no. 22, April 1970.

55 GREENWOOD, J., *Trent & Humber Steam Packet Companion*, 1833.

56 HAWTHORNE, N., *Our Old Home*, *c.*1870.

57 GRESWELL, REV. J., *An Account of Runcorn and its Environs*, *c.*1809.

58 WALFORD, E., *Old and New London*, *c.*1876.

59 WELLS, H. G., *History of Mr Polly*, 1910.

60 THOMAS, N. L., *The Story of Swansea's Districts and Villages*, 1969.

61 DAVIES, G. C., *The Handbook to the Rivers & Broads of Norfolk and Suffolk*, 1907.

62 BONTHRON, P., *My Holidays on Inland Waterways*, 1916.

63 British Waterways Board, '*The Facts About the Waterways*', 1965.

64 Pamphlet published by local landowners, 8 February 1825, Essex Record Office D/DC2 E2.

Index of Canals and Navigations

General Index